W9-BEI-437

Synthetic Pyrethroids

Michael Elliott, EDITOR
Rothamsted Experimental Station

A symposium sponsored by the Division of Pesticide Chemistry at the 172nd Meeting of the American Chemical Society, San Francisco, Calif., Aug 30–31, 1976.

ACS SYMPOSIUM SERIES **42**

AMERICAN CHEMICAL SOCIETY
WASHINGTON, D. C. 1977

Library of Congress CIP Data

Synthetic pyrethroids.
(ACS symposium series; 42 ISSN 0097-6156)

Bibliography: p.
Includes index.

1. Pyrethroids—Congresses. 2. Insecticides—Congresses.
3. Chemistry, Organic—Synthesis—Congresses.
I. Elliott, Michael, 1924- II. Series: American Chemical Society. ACS symposium series; 42.

SB952.P88S96 632'.951 77-1810
ISBN 0-8412-0368-7

PRINTED IN THE UNITED STATES OF AMERICA

ACS Symposium Series

Robert F. Gould, *Editor*

FOREWORD

The ACS Symposium Series was founded in 1974 to provide a medium for publishing symposia quickly in book form. The format of the Series parallels that of the continuing Advances in Chemistry Series except that in order to save time the papers are not typeset but are reproduced as they are submitted by the authors in camera-ready form. As a further means of saving time, the papers are not edited or reviewed except by the symposium chairman, who becomes editor of the book. Papers published in the ACS Symposium Series are original contributions not published elsewhere in whole or major part and include reports of research as well as reviews since symposia may embrace both types of presentation.

CONTENTS

PREFACE

The valuable insecticidal properties of pyrethrum were recognized in the 19th century and stimulated detailed examination of the chemical constitution of the active esters in the first quarter of the 20th century. Although the acidic components of the esters were correctly identified at an early stage in these studies, only in 1947 were the structures of the alcohols settled. The first synthetic pyrethroid, allethrin—still important today—was developed soon afterwards. By 1968, tetramethrin, a good knockdown agent, and resmethrin and bioresmethrin, the first synthetic compounds with greater insecticidal activity and lower mammalian toxicity than the natural esters, had been discovered. These compounds did not greatly extend the range of application of pyrethroids for, like the natural compounds, they were unstable. Investigation of structure–activity relationships continued, and by 1973 compounds had been developed which were more photostable yet retained many of the favorable characteristics of the natural esters and earlier synthetic compounds. The new materials are now being critically assessed to establish those applications where their special combination of properties may be especially advantageous; in some instances, they may replace existing insecticides which have properties no longer considered acceptable.

The structures of pyrethroids are more complex than those of other major classes of insecticides, and they are relatively expensive to manufacture; however, their greater insecticidal activity, permitting fewer applications of lower doses, may give an advantage over present compounds, especially when persistent residues in the environment must be avoided. Further, industrial processes have been developed by which the most active optical and geometrical isomers of some pyrethroids could be prepared on a multi-ton scale—an outstanding achievement of modern chemical technology.

The introductory paper on "Synthetic Pyrethroids" reviews the compounds now available from an historical viewpoint and traces development of our understanding of relationships between chemical structure and insecticidal activity, photostability and mammalian toxicity. The general implications of the discovery of more stable compounds, which may be considered to constitute a new group of insecticides, are important themes of this collection of papers.

Pyrethroids are flexible molecules and their conformations probably greatly influence their insecticidal activity. In the first paper, preferred conformations are calculated solely on the basis of non-bonded interactions and considered in relation to the conformations in the solid state and to biological results. The next papers consider the effects of modifying the acidic and alcoholic components of pyrethroids; nearly all the new compounds are less active than the parent esters on which they are based. These results and other work reviewed in the book show the difficulty of discovering new acidic and alcoholic components of synthetic pyrethroids with the combination of properties necessary to justify commercial development. The most promising examples so far are 3-phenoxybenzyl and α-cyano-3-phenoxybenzyl alcohols and the optical and geometrical isomers of 3-(2,2-dihalovinyl)-2,2-dimethylcyclopropanecarboxylic acids and α-(4-chlorophenyl)isovaleric acid. Much of the volume therefore describes syntheses, the biological properties, metabolism, and analysis of esters formed from combination of these components.

Although recognized as nerve poisons, the detailed mechanism by which pyrethroids act remains obscure; however, investigations described here on the housefly nervous system and on the crayfish abdominal nerve cord disclose many significant results. The observation that some compounds which are potent insecticides have relatively weak action on the nervous system of crayfish compared with closely related esters which are much less active insecticidally may have important implications.

The final group of papers deals with factors concerning the practical application of pyrethroids. Although there is now abundant evidence that pyrethroids with appropriate structures are sufficiently stable to control agricultural pests, it is important to establish precisely how long they persist and the nature and toxicity of their metabolites in various environments. The results so far indicate that although some of the newer compounds are relatively photostable, they are readily metabolized by organisms which have esteratic or oxidative mechanisms to non-toxic products which do not accumulate in mammalian systems. The papers presented thus span the many rapidly advancing aspects of pyrethroid studies and indicate that some of the newer pyrethroids discussed may make valuable, practical contributions to insect control within a short time.

The symposium "Synthetic Pyrethroids: Recent Advances" arranged by the Division of Pesticide Chemistry of the American Chemical Society at the 1976 Autumn meeting in San Francisco was therefore particularly opportune. It was complemented by a meeting of the Pesticides Group of the Society of the Chemical Industry in London in November 1976 on "Newer Applications of Pyrethroids."*

The Division of Pesticide Chemistry chose to honor me at this time with the Burdick and Jackson International Award for Pesticide Chem-

* *Pestic. Sci.* (1977) 8 (in press).

istry. I am conscious of the prestige of the Award, and it is noteworthy that the Division should make this characteristically warm and generous gesture to a British chemist in the 100th year of the American Chemical Society and at the time of the Bicentennial celebrations of the Declaration of American Independence. It is also appropriate to recognize and to acknowledge the broad base of international research on which the work of my colleagues and myself rests and into which it is integrated.

The Award having been given for work on pyrethroid insecticides, it is fitting to recall the wisdom which led F. B. LaForge in the United States and Frederick Tattersfield and Charles Potter, successive heads of the Insecticides and Fungicides Department, Rothamsted Experimental Station, to continue to investigate the insecticidal action and chemical properties of pyrethrum. Potter at Rothamsted and Stanley Harper at the Universities of Southampton and London discerned the long-term advantages of continuing to study this relatively complex group of non-persistent insecticides with low mammalian toxicity even when immediate applications appeared limited by the development of major groups of synthetic insecticides.

I owe a considerable debt to Stanley Harper and Charles Potter for help and support over a long period and more recently to Norman Janes, whose scientific and personal cooperation has been of rare and outstanding quality. David Pulman has contributed greatly to our work by his skill and perseverance. We thank Roman Sawicki, Paul Needham, and Andrew Farnham for many bioassay results, essential to our progress, and many other colleagues for valuable help and discussions.

Harpenden, Herts., England
December 1976

MICHAEL ELLIOTT

INTRODUCTION

The following remarks are those of Professor Ryo Yamamoto, Professor Emeritus of the Tokyo University of Agriculture. He represents the many Japanese chemists who have made distinguished contributions to the knowledge of natural and synthetic pyrethroids. Professor Yamamoto was investigating the structure of the pyrethrins in Japan during the period when Staudinger and Ruzicka were working in Switzerland.

It is an honor and a great pleasure for me to introduce Dr. Elliott's award collection "Synthetic Pyrethroids" and to offer my congratulations.

I am an old chemist. It was 1923 when I first derived *trans*-caronic acid from the natural pyrethrins and demonstrated the presence of the cyclopropane structure in the chrysanthemic acid moiety in Tokyo. Now in 1976, I am an active member of Pesticide Science Society of Japan and still interested particularly in the science of pyrethroids.

I am deeply impressed by the development of pyrethroid chemistry: from structural assignment of natural pyrethrins to recent developments of synthetic pyrethroids. These are all puzzling, and I can hear the early rumblings of what may become "Pyrethroid Age." I am very pleased to learn that permethrin by Dr. Elliott and S-5602 by the Sumitomo group are particularly promising for agricultural uses and those interested in pyrethroids are developing newer and newer ideas. Further development of pyrethroids will be accelerated not only by studying the chemistry but also by elucidating the biological aspects, particularly the mode of action. Here is a wonderful area of research and development for all.

San Francisco, Calif.
August 1976

RYO YAMAMOTO

1

Synthetic Pyrethroids

MICHAEL ELLIOTT

Rothamsted Experiment Station, Harpenden, Hertfordshire, AL5 2JQ, England

Insecticides with a range of physical, chemical and biological properties will be required for as long as present methods of crop protection continue and until diseases transmitted by insects no longer affect man and his livestock. Ideally, both established and new products will be used efficiently (1,2) in rationally conceived pest management schemes, (3,4) in some cases complemented by new approaches to insect control. (5,6,7,8). Millions of human beings owe their freedom from starvation and protection from disease to insecticides. Nevertheless, the present range of compounds is inadequate because resistant insect species have emerged to diminish their effectiveness for some applications, because they have been judged unduly persistent or excessively toxic to men and mammals or because they are not sufficiently selective between pests and beneficial insects. New insecticides with superior properties are needed; to indicate what improvements might be possible some of the physical and biological properties of the classes of insecticides at present available will first be reviewed.

Table I - Properties of Classes of Insecticides

Class	Polarity, Log P*	Approximate Solubility in water, p.p.m.	Systemic Action
Carbamates	-1 to 3	> 40	+ and -
Organophosphates	1 to 5.5	> 1	+ and -
Organochlorines	5.5 to 7.5	< 1	-
Pyrethroids	4 to 9	< 1	-

* P = Octanol-water partition coefficient

Table 1 shows that most carbamates (9) and many organophosphates (10) are relatively polar, water-soluble compounds, a number of which have useful systemic and translaminar properties. In contrast, most organochlorine insecticides are non-polar, stable and therefore relatively persistent compounds (11,12). The natural pyrethrins (12,13,14,15,16,17,18,19) and all the synthetic pyrethroids used at present are also non-polar compounds, as indicated by their octanol-water partition coefficients (20,21), and have very small solubility in water. They also, therefore, have no systemic or translaminar properties. Unlike the organochlorine compounds, however, they are unstable and non-persistent, restricted in their applications by these characteristics and because they are more complex and more expensive to produce than the other three groups of insecticides. New insecticides should combine, as far as possible, the most valuable properties of these groups.

Table II - Toxicities of Classes of Insecticides to Insects and Mammals[a]

Class	Rats[b]		Insects[c]		Ratio
Carbamate	45 $mg.kg^{-1}$	(15)	2.8 $mg.kg^{-1}$	(27)	16
Organophosphate	67 " "	(83)	2.0 " "	(50)	33
Organochlorine	230 " "	(21)	2.6 " "	(26)	91
Pyrethroid	2000 " "	(11)	0.45" "	(35)	4500

[a] Geometric means of no. of data items in brackets
[b] From published acute oral LD50 values
[c] From published values, principally to 4 species, by topical application.

Simplified biological data for the four groups of insecticides in Table II demonstrate a relative advantage for pyrethroids. The level of insecticidal activity attainable with carbamates, organophosphates and organochlorine compounds is remarkably similar (22,23) and apart from a few special cases (24) intensive research over three decades (25) has failed to discover acceptable compounds in these categories with generally greater potency to a wide range of species. In contrast, investigation of pyrethroids during the same period, with much smaller total research effort, has revealed compounds with progressively increasing activity, up to four or five times higher than that of the other classes to most insect species (26).

Relative safety is indicated by the ratio of toxicities to rat and insect (Table II, column 4). In this respect pyrethroids are also clearly superior, because they are both very active against insects and relatively non-toxic to mammals.

The scope for structural variation in pyrethroids and the restricted attention given to them suggest that detailed knowledge of the chemical and biochemical basis for their insecticidal action might show how related compounds with improved properties could be discovered. Therefore progress in research and development in this challenging area is reviewed in this and subsequent contributions to the symposium.

Structure and Activity of Pyrethroids

The evolution of synthetic pyrethroids can be assessed appropriately by relating their activity to that of pyrethrin I which has an LD50 of 0.33ug per female house-fly (27) and provides a convenient prototype and standard. In the figures in this paper, successive filled boxes show ten fold changes in activity relative to pyrethrin I, compounds less active being on the left of the arrow and those ten, one hundred and one thousand times more active than this standard to the right. Thus pyrethrin II, which with pyrethrin I is the most important constituent of natural pyrethrum (18,19,28) and decamethrin, with LD50 values per house-fly of 0.20ug (29) and 0.0003ug (30), respectively are represented as shown in Figure 1.

The structures of pyrethrin I (especially effective for kill (31)), of pyrethrin II (a good knockdown agent) and of decamethrin illustrate features required for highest activity in this group of insecticides. All three compounds are cyclopropane carboxylic acid esters with two methyl groups on C-2 and an unsaturated side chain on C-3, *trans* to the carboxyl group in the natural esters, *cis* in the synthetic compound. The relative disposition in space of substituents at the carboxylic acid centre, C-1, is important, compounds of the opposite stereochemical configuration, (S), being much less active (32); (a nomenclature appropriate for this series is discussed by Elliott *et al.*, (33)). The acid in each ester is combined with a secondary alcohol of which the hydroxyl group is either part of a nearly planar cyclopentenolone ring (34) as in pyrethrins I and II or is attached through a tetrahedral carbon atom to an aromatic ring. A centre of unsaturation (*cis*-butadienyl, or phenyl) is linked *via* a methylene or

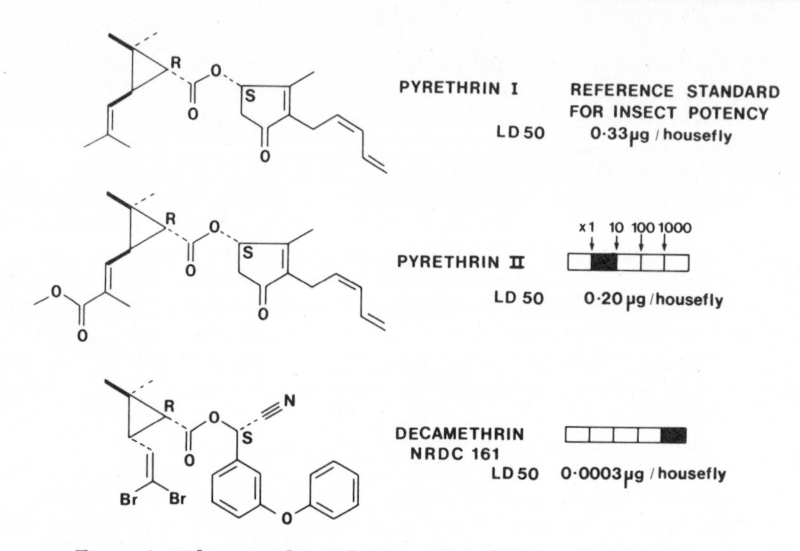

Figure 1. The natural pyrethrins compared with a synthetic ester

oxygen bridge and a 3 or 4 carbon spacing unit to the chiral centre which bears the hydroxyl group.

From their work in the years 1910-1916 (published in 1924 (35)) it is clear that Staudinger and Ruzicka, who, like R. Yamamoto (36,37,38), were eminent pioneers of pyrethrum chemistry, recognised many of the features in the structures of pyrethroids necessary for insecticidal activity. They knew that the constituent alcohols and acids were only active when combined with one another, and that an intact ester-linkage was important. The gem-dimethyl group on the cyclopropane ring was required and activity diminished by saturating the side chains in acidic and alcoholic components. Staudinger and Ruzicka detected insecticidal activity in the chrysanthemate of the cyclopentenolone (Figure 2) which had an allyl rather than a pentadienyl side chain. This concept of a shorter unsaturated side chain was later used by Schechter, Green and La Forge (39,40,41,42) when they developed allethrin, the first important synthetic pyrethroid. Dienic unsaturation in the alcoholic side chain was therefore not necessary to attain a practical level of insecticidal activity.

Staudinger and Ruzicka also detected insecticidal activity in esters of piperonyl alcohol (Figure 2) (as later did Synerholm (43)) and 4-isopropylbenzyl (cuminyl) alcohol, foreshadowing the important series of benzyl esters such as those of 3-phenoxybenzyl alcohol (31,44) in modern pyrethroids. They further examined compounds incorporating open chain equivalents of cyclopropane carboxylates, envisaging in principle the non-cyclic compounds which Ohno and his co-workers in 1974 (45,46) demonstrated to be important insecticides; the gem-dimethyl group of the cyclopropane was retained as isopropyl, with an unsaturated centre placed on the α-carbon atom of the acid.

Development of the relatively complex basic chemistry of the natural pyrethrins and related synthetic compounds was greatly assisted by spectroscopic methods, ultraviolet (47,48) and infrared initially (49) and more recently nuclear magnetic resonance (50) and mass spectrometric (51) techniques, without which investigation of the viscous, unstable, high-boiling liquids and their isomerization (52,53,54) and degradation products (55) was relatively difficult. Until

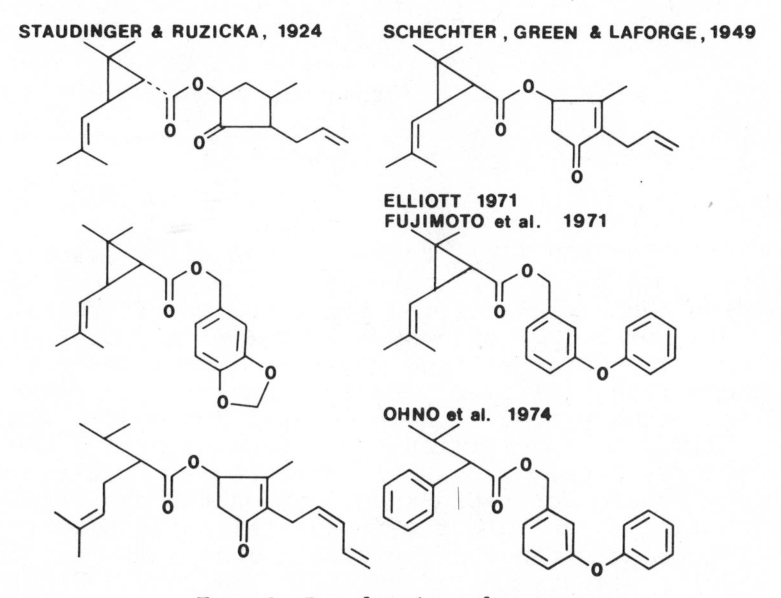

Figure 2. Precedents for modern structures

ten years ago only one synthetic pyrethroid, allethrin, was significant in practice, but since then progress in developing new compounds has been rapid, and, as the basis for further discussion, active compounds now available will be reviewed.

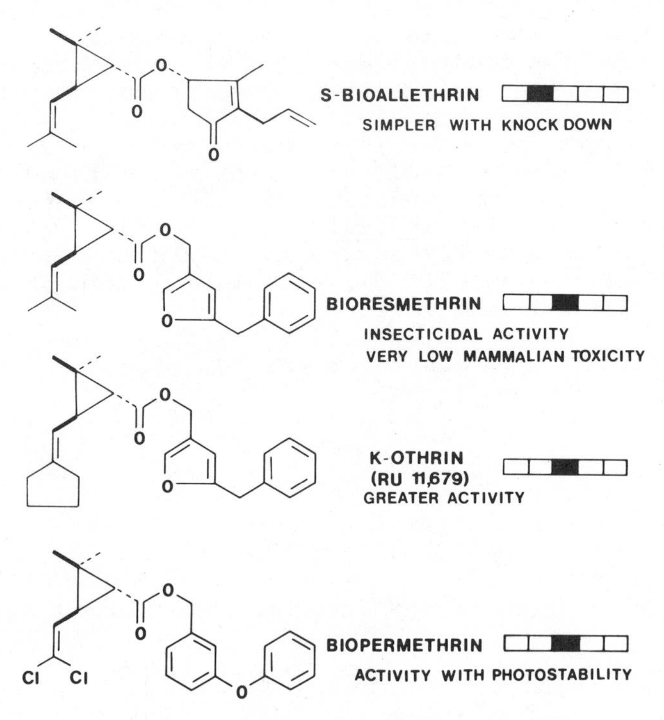

Figure 3. Developments of the basic structure

S-Bioallethrin (56), Figure 3, has all the structural features of natural pyrethrin I, except that an allyl side chain replaces the cis-pentadienyl system. It is more polar (20) than pyrethrin I with faster knockdown but poorer killing power to most insects except house-flies (31,57). Early bioassays overemphasised the potency of allethrin because frequently they were confined to house-flies and evaluated mainly the knockdown response; the subtle differences in basic structure needed for rapid knockdown on the one hand and high kill on the other were not recognised (20). The outstanding potency of pyrethrin I against many insect species was overlooked until it was obtained pure (55) and undiluted by less active components (28).

Continued investigation of the structural requirements for high insecticidal activity in pyrethroids led to the compound bioresmethrin (58,59) (Figure 3), in which the cyclopentenolone nucleus has been replaced by the sterically equivalent furan ring, and the unsaturated alkenyl side chain by an aromatic nucleus. Bioresmethrin was the first synthetic pyrethroid to show equal or higher killing activity than the natural compounds against many insect species (26,31,58) and, a welcome property unanticipated during the developmental work, lower mammalian toxicity (60).

With the same furan alcohol, but the more lipophilic ethanochrysanthemic acid in the compound K-Othrin (R/U 11,679) (61,62,63) Martel and co-workers increased insecticidal activity still more (31) although at the expense of higher mammalian toxicity.

Insecticidal activity of cyclopropanecarboxylates was raised even more by another modification at the same site in the molecule: substituting chlorine for the methyl groups in the isobutenyl side chain (64,65, 66). This transformation also had the important consequence of eliminating the principal photosensitive centre in the acid (67,68). Correspondingly, replacing the photolabile 5-benzyl-3-furylmethyl unit or other previous alcohols with 3-phenoxybenzyl gave the compound biopermethrin (67) the first synthetic pyrethroid with adequate stability for field use. Insecticidal activity is maintained on a leaf surface for two weeks or more in bright sunlight, without unduly long persistence in the soil (69,70). Biopermethrin also retained the low oral and intravenous mammalian toxicity of the unstable synthetic pyrethroids (67,71).

Conformation and Activity of Pyrethroids

The natural pyrethrins and the synthetic compounds just reviewed are all flexible molecules. In the light of present knowledge their insecticidal action is best interpreted as an ability to adopt a conformation in which all the structural features essential for potency are appropriately oriented with respect to each other and to a complementary receptor. A characteristic of pyrethroids is the sensitivity of their insecticidal action to changes in substituents at certain important centres by which either the balance of conformers present is disturbed, or contact of the molecule with a receptor is obstructed. Valuable indications of the characteristics of the receptor and conformations needed for optimum insecticidal activity can be gained by the following detailed examination of the sensitive positions in the molecule.

One such important site is the chiral centre in the acid component to which the carboxyl group is attached (Figure 4). Esters of cyclopropane carboxylic acid with substituents in the (R) configuration shown, whether the side chain is trans or cis to the carboxyl group, are active, whereas esters of the (S) epimers are inactive, or much less active. Similarly (S)-isopropylaryl acetates, which correspond to (IR)-chrysanthemates in their chiral arrangement of substituents (45,46,72), are much more active than their (R) epimers. This is strong evidence that interaction with a chiral receptor is involved in the lethal action, since in all phenomena involving migration and partition, for example at a phase boundary, each member of a pair of isomers will behave identically.

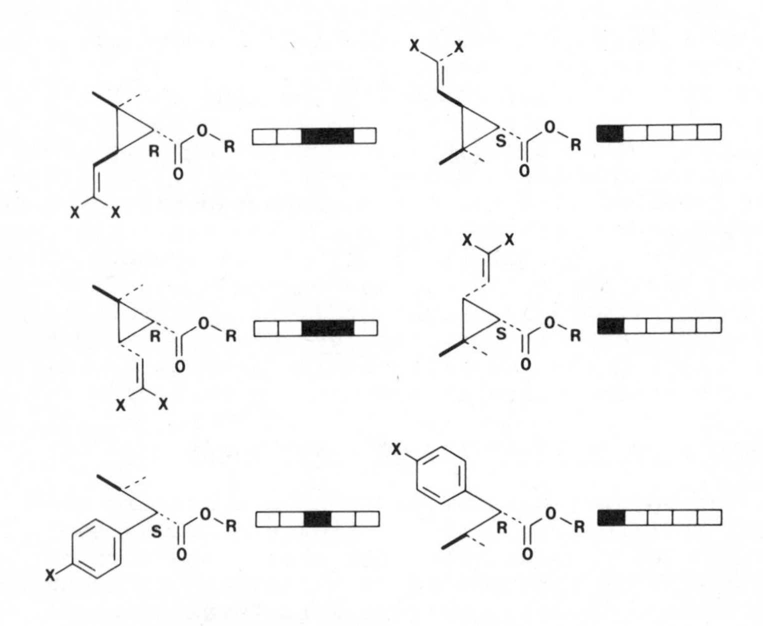

Figure 4. Potency of esters from [R]- and [S]-forms of cyclopropane and phenylacetic acids

The potency of esters of cyclopropanecarboxylic acids is also sensitive to substitition at or on the side chain at C-3 (Figure 5). Some compounds with a trans-dichlorovinyl substituent (see, for example, compounds in Figure 4, x = Cl) are powerful insecticides. An additional methyl substituent greatly diminishes potency; the activity of alkenyl cyclopropanecarboxylates (Figure 5) (73,74) is also lowered by a 1-methyl substituent. An analogous depression of activity by methyl substitution occurs in the α-isopropylaryl acetates of Ohno et al (45,46), where ortho-substituted aryl compounds are much less active. In the three series of compounds in Figure 5 the added methyl groups are at sites in the molecule where they may disturb preferred conformations, as discussed in the succeeding paper (74) or may block access to an essential position on the receptor site.

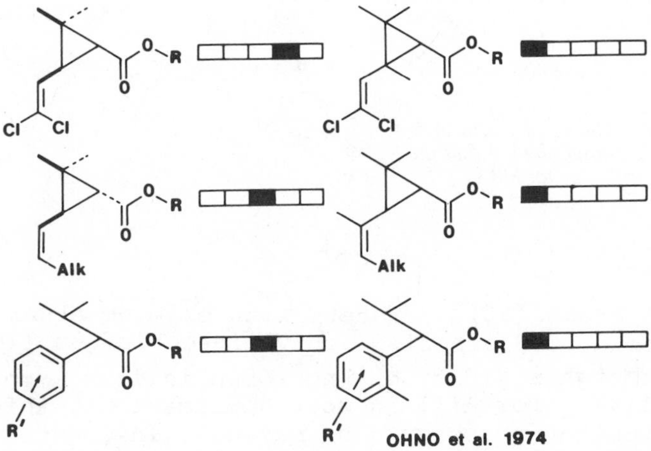

OHNO et al. 1974

Figure 5. Influence of methyl substituents on acid components of various esters

These examples show that the acid side chain attached at C-3 of the cyclopropane ring is a position where structural changes greatly influence insecticidal activity. In the variations shown in Figure 6, activity again depends on the nature of the substituent at this site. Thus, if there are no methyl groups at C-3 or C-1 (cf. Figure 5) extremely high insecticidal activity is attained in esters with Z- and E-butadienyl and -pentadienyl substituents trans, and, to a smaller extent, cis to the (1R) carboxyl centre (33,75). Further, some esters of 3-dihalovinyl substituted acids are outstandingly potent insecticides (30,64,65); in this series, the cis esters are usually more active

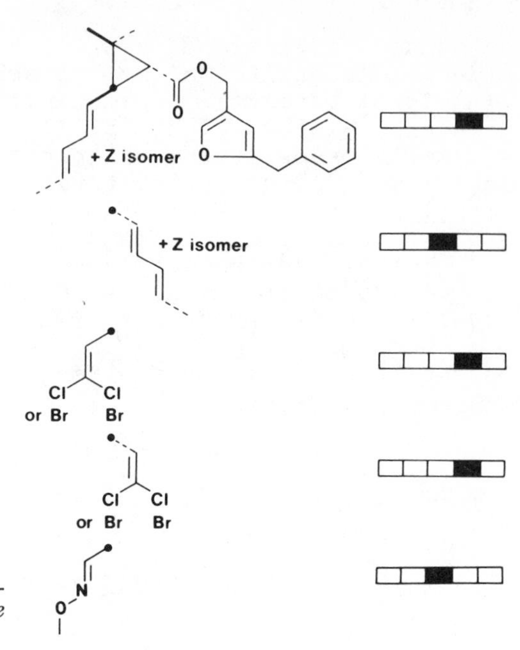

Figure 6. Side chains of esters of 5-benzyl-3-furyl methyl alcohol that are very effective for kill

than the trans (66). Esters with oximino ether substituents at C-3 are also active insecticides (76).

Although rapidity of knockdown is less important than activity for kill in most applications, this is an interesting property also markedly influenced by modifications of the C-3 substituent. In Figure 7, relative knockdown efficiency is indicated rather than kill. The pyrethrate related to bioresmethrin is a better knockdown agent than bioresmethrin itself (20), while the difluorovinyl compound (NRDC 173) (66) is even more active in this respect. The thiolactone, Kadethrin, R/U 15,525 acts more rapidly against houseflies than any other compound yet reported (77). The delicate balance between structure and activity in pyrethroids is demonstrated by the fact that the related 3-phenoxybenzyl ester (73) almost completely lacks knockdown activity. In most compounds, as with Kadethrin, good knockdown is only achieved at the expense of killing activity. However, the difluorovinyl ester (Figure 7) is an exception in this respect, because it combines good knockdown action with killing power three times as great as that of bioresmethrin (66,78,79).

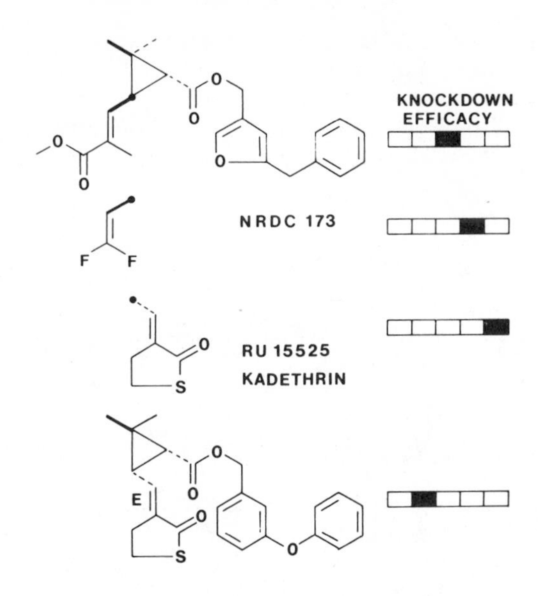

Figure 7. Modification of acid side chain for knockdown

Like cyclopropanecarboxylates, the activity of the α-isopropylaryl acetates introduced recently by Ohno and co-workers (45,46) is very sensitive to structure and substitution. The dichloroisostere of the isopropyl compound (Figure 8, R = 3-phenoxybenzyl or α-cyano-3-phenoxybenzyl) is inactive, possibly because hydrogen chloride is eliminated extremely rapidly to give a monochlorolefin lacking the structural characteristics for insecticidal action. The isosteric amine and carbamate (73) are also inactive.

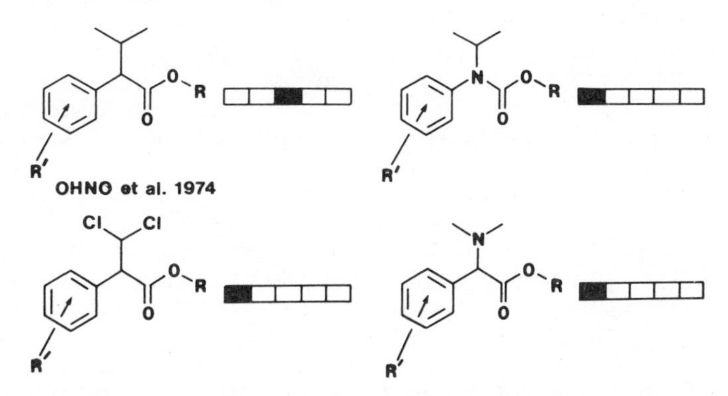

Figure 8. Compounds related to α-isopropylphenyl acetates I

Ohno et al. (45,46) also showed that ethyl substituted compounds had insecticidal activity only slightly below that of the isopropyl derivatives (Figure 9). However, both the related diethyl and monomethoxy compounds shown were inactive (73).

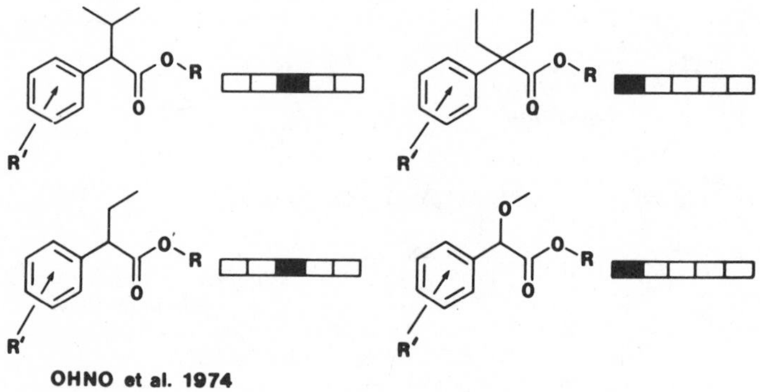

OHNO et al. 1974

Figure 9. Compounds related to α-isopropylphenyl acetates II

In another series of variations examined (73) (Figure 10) the aromatic centre was displaced by an oxygen or methylene bridge to a position more remote from the chiral centre. The compounds were not active. Two further compounds, one a phenylcyclopropane, the other a tetrahydronaphthalene in which the isopropyl group was locked in either of two ways were also non-toxic; they show that in both cyclopropanecarboxylates and isopropylarylacetates a precise structure in the appropriate configuration is needed for insecticidal activity. These additional results support the conclusion (45) that the insecticidal activity of both cyclopropane carboxylates and α-isopropylarylacetates depends on common structural features.

Such examples illustrate how greatly insecticidal activity is influenced by small changes in the structure of acidic components of pyrethroids. The alcoholic constituents are equally sensitive, as compounds substituted at the α-methylene groups of esters of furfuryl, furylmethyl, and benzyl alcohols exemplify (in Figure II R is a representative cyclopropanecarboxylate with R' = H, Me or CN).

X = CH_2 or O

Figure 10. Compounds related to α-isopropylphenyl acetate III

Esters of 5-benzyl-3-furylmethyl alcohol (R' = H) are usually two to three times as potent as those of 3-phenoxybenzyl alcohol and some ten times more potent than those of 5-benzylfurfuryl alcohol. A methyl substituent (R' = Me) almost eliminates activity of a 3,5-disubstituted furan ester and depresses that of 3-phenoxybenzyl derivatives. Esters of α-cyanoalcohols (R' = CN) are most interesting. The cyano substituent has little influence on the activity of 2,5-disubstituted furan derivatives, depresses that of 3,5-furans

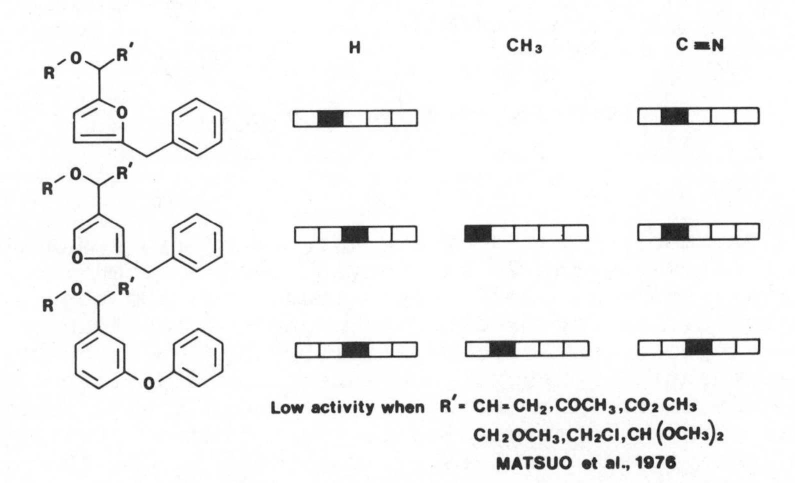

Figure 11. Influence of substituents at the α-methylene group

and increases the activity of 3-phenoxybenzyl esters to produce the most effective alcoholic constituent for pyrethroids yet reported (30,66,80). Other more bulky substituents at this site in the molecule depress or eliminate activity (80).

The potency of esters of (±)-α-cyano-3-phenoxybenzyl alcohol stimulated attempts to isolate pure isomers. The mixed esters (NRDC 156) of the (1R)-cis-dibromovinyl acid in hexane gave crystals (m.p.100°; NRDC 161, decamethrin) and a liquid still containing some NRDC 161 (30). The crystalline isomer was estimated to be about six times as toxic as the liquid, itself little more active than the 3-phenoxy-benzyl ester (32). Whether significant or coincidental, it is notable that the esters of the two optical forms of allethrolone also differed six fold in activity (Figure 12) (81,82) although at a level of potency to houseflies some six hundred times lower.

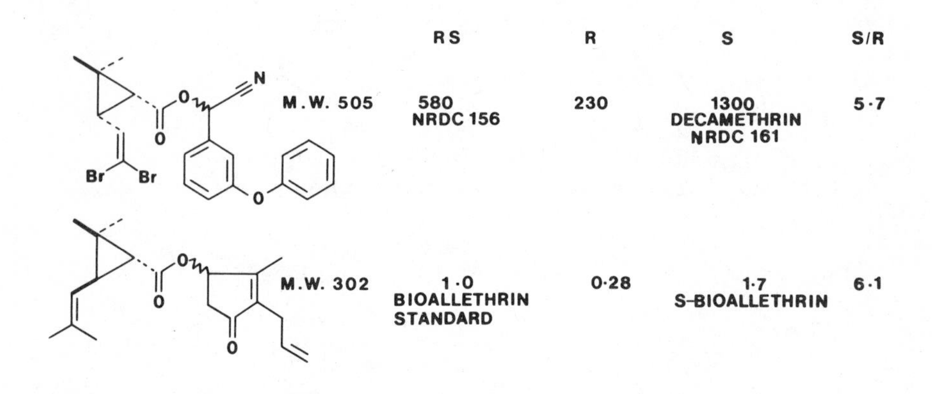

Figure 12. Relative potencies of esters of chiral alcohols

Because the ester of the (S)-cyanhydrin had such great activity, the 2- and 6-cyano-3-phenoxybenzyl esters (Figure 13) were synthesised, with the possibility that in one or other compound the location of the α-cyano group would be simulated by the extra cyano substituent. However, both esters were inactive (73).

The ester link (Figure 14) is another site where small changes in structure greatly influence activity. The substituted phenyl ester (top) which lacks the methylene group of the benzyl esters is inactive (73) and although, as noted, some cyanohydrin esters

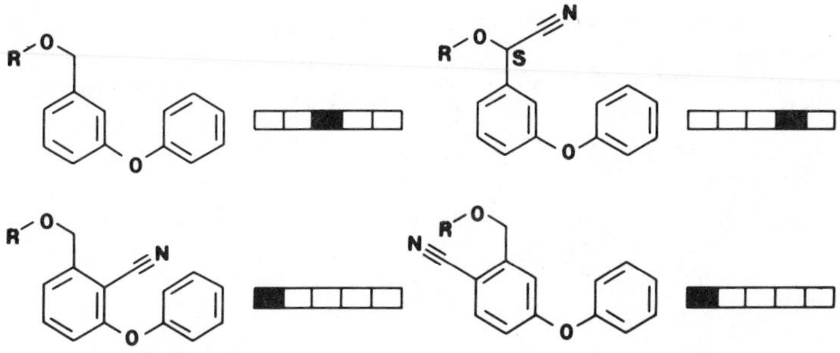

Figure 13. α-, 2-, and 6-cyano-3-phenoxybenzyl esters

(centre) are potent, the corresponding cyano-amides are not. A common factor in these modifications may be the changed conformations induced at this central link, which would greatly influence the overall shape of the molecule.

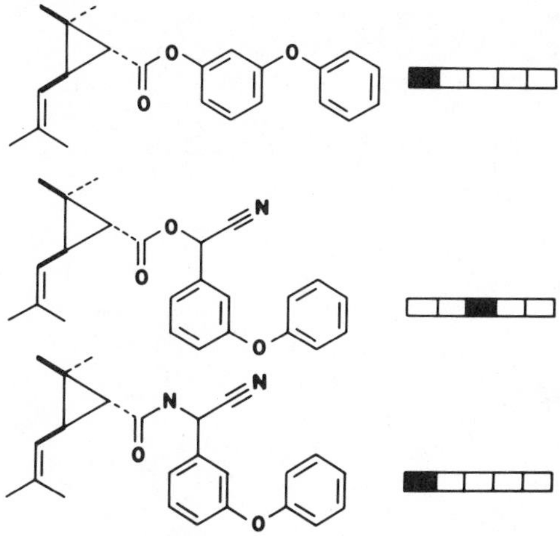

Figure 14. Modifications at the ester bond

Some unsaturation is present in the side chain of the alcoholic component of all powerful pyrethroids described so far, but again small changes remove activity (Figure 15). Thermal isomerisation of pyrethrin I to a compound with the trans-cis dienic system conjugated with the ring, as shown (52) almost eliminates activity, and although the benzyl furan bioresmethrin is insecticidal, the corresponding phenylfuran is

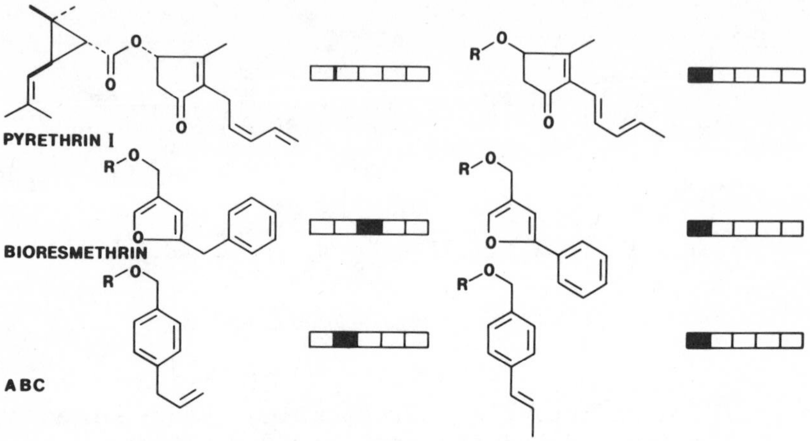

PYRETHRIN I

BIORESMETHRIN

ABC

Figure 15. Esters of alcohols with various side chains

again almost inactive (83,84). The 4-propenylbenzyl chrysanthemate is much less active than the 4-allyl compound (85,86). The location of unsaturation with respect to the nucleus is therefore important, shown also by the inactivity of the compounds in Figure 16, which are related to 3-phenoxybenzyl esters, but have an extra methylene group.

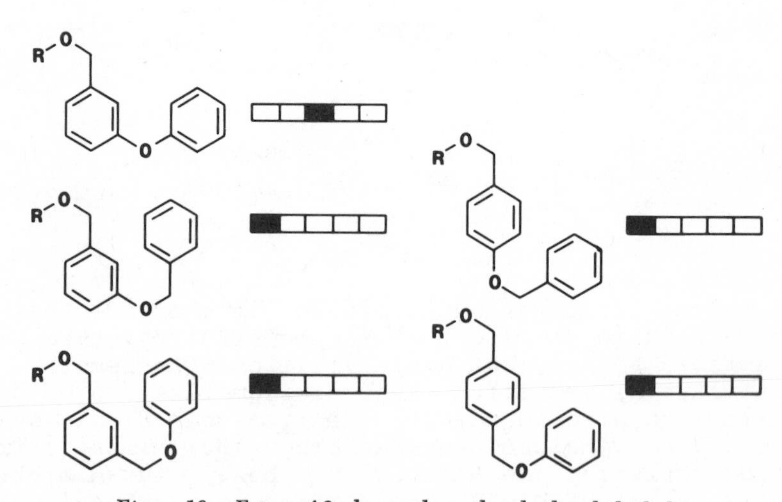

Figure 16. Esters of 3-phenoxybenzyl and related alcohols

The field of synthetic compounds with structures related to the natural pyrethrins is developing very rapidly and may perhaps be considered in a state comparable to that of organophosphates and carbamates some twenty years ago. Therefore modification of some of the structure activity principles discussed may be necessary with fresh discoveries. However, present knowledge reviewed here is consistent with the conclusion indicated earlier that the insecticidal action of these compounds requires a number of sites on the molecular framework to be appropriately oriented so that intimate contact with a complementary chiral surface or structure is possible. The probability that this particular orientation will be achieved may depend on the balance of conformers in which the molecule exists.

Such concepts provide a reasonable, though at present necessarily superficial, interpretation of the sensitivity of the insecticidal action of pyrethroids to small changes of substituent or configuration at certain centres; it implies that their lethal action involves the intact molecules, unlike that of organophosphates and carbamates where cholinesterase is phosphorylated or carbamoylated (87,88) in a biochemical reaction in which the nature of the leaving group is not critical.

An exciting question, to which further research may provide an answer, is whether the molecule is in the fully active conformation at the moment of first contact, or forms a bound complex by a sequence of steps following initial impact at one of the active centres - the so-called "zipper" concept (89).

Photostability and Mammalian Toxicity

This survey of synthetic pyrethroids has shown the rapid advance in knowledge of the relationship of insecticidal activity to structure, contributing to fundamental understanding of biochemical processes in insects. Two other regions of research influence the development of synthetic pyrethroids for practical application in the field. These are the relationships of structure to photostability and to mammalian toxicity.

For many years, it was recognised that if the natural pyrethrins and allethrin could be induced to persist on leaf surfaces in sunlight, they might be very valuable for controlling agricultural crop pests, especially if their other favourable characteristics could be retained. However, the natural pyrethrins are notoriously unstable and attempts to protect them from UV light with filters (90), as inclusion compounds (91) and otherwise are only partly successful, particularly as the pyrethrins are already more expensive than competing insecticides. Recently, in a more direct approach photostable pyrethroids have been evolved, (26,45,46,67,92,93) which may assume an important role in insect control in the future.

Figure 17 shows some stages in the development of pyrethroids stable enough to use in the field. The figures in brackets are insecticidal potencies against Anopheles stephensi relative to DDT (70,94). The asterisks show photolabile sites in the molecules.

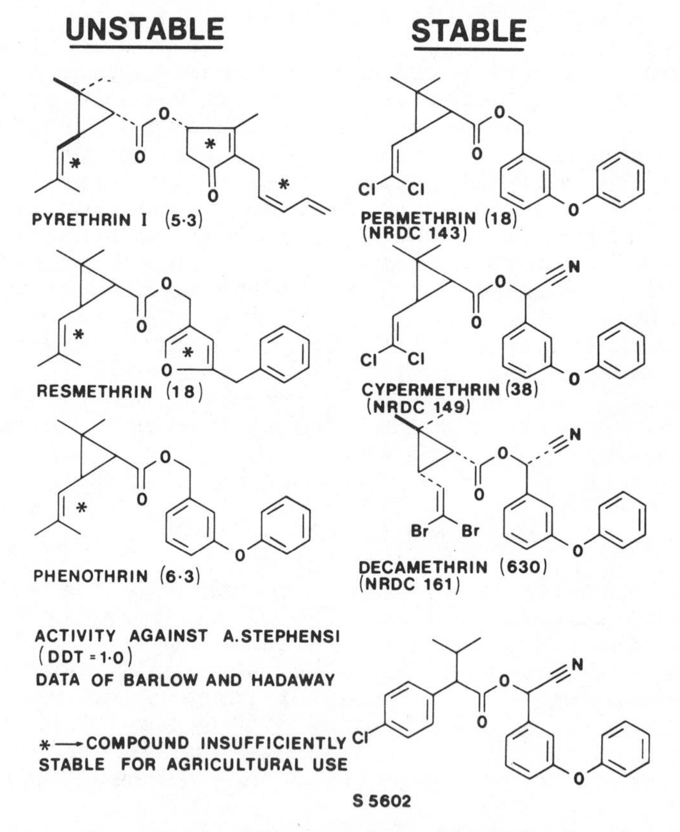

Figure 17. Development of photostable pyrethroids

Investigations with pure pyrethrin I (55,95) and related compounds (68) identified centres in the structures sensitive to photosensitised attack by oxygen, the more important ones being the side chain of chrysanthemic acid and the cis-pentadienyl side chain. Systematic synthesis and testing (59) of a series of compounds related to pyrethrin I led to resmethrin, a more powerful insecticide in which a stable aromatic ring had been substituted for the photosensitive diene side chain but which had a photolabile furan ring (67, 96). A further step was the recognition that a meta substituted benzene ring was equivalent in some respects to the 3,5-furan in resmethrin (31) and in the compound phenothrin (44) where the methylene bridge is also replaced by an oxygen link, the alcoholic component is photostable. However, phenothrin still contains the labile chrysanthemate centre, so is not sufficiently stable for most agricultural applications and, moreover, is generally less active than resmethrin. In permethrin a dichlorovinyl side chain replaces the isobutenyl unit of chrysanthemic acid with enhancement of insecticidal activity and, all photolabile centres having been eliminated, the compound is more stable on leaf surfaces than many organophosphates and carbamates. Nonetheless, when exposed to systems active in metabolism of organic compounds, for example, microorganisms in the soil, it is degraded sufficiently rapidly to allay any concern about undue accumulation.

The α-cyano group in cypermethrin gives still greater insecticidal activity, albeit with somewhat increased mammalian toxicity. As described, invstigating the combinations of optical and geometrical isomers of the dihalovinyl acids with the cyanohydrins led to the discovery of the outstandingly potent compound decamethrin (NRDC 161) which apparently has each centre in the optimum configuration for activity. Decamethrin (30,32) which is also adequately stable (97,98) for field use (70,94,99) is sixteen times as active as cypermethrin and six hundred times as active as DDT to A. Stephensi (94) and is the most powerful lipophilic insecticide yet synthesised. A further important stage in the evolution of pyrethroids for use in agriculture was the discovery of the activity of esters of α-cyano-3-phenoxybenzyl alcohol with non-cyclopropane acids such as α-isopropyl-4-chlorophenylacetic acid. Most information has so far been published on fenvalerate (sumicidin, S 5602) (45,46,70).

The other important area of investigation which influences development and practical application of synthetic pyrethroids is the molecular basis for mammalian toxicity, to which important contributions have been made by groups led by J.E. Casida in the University of California at Berkeley and by J. Miyamoto of the Sumitomo Chemical Company. The topic is covered in more detail later in this symposium and here the discussion is restricted to some emerging principles.

The oral toxicities to mammals of many synthetic pyrethroids are so low (60,71) (8000-10,000 $mg.kg.^{-1}$ for female rats) that comparing them gives little guidance to the structural factors influencing mammalian toxicity. Intravenous toxicities to female rats, usually some ten times greater than the oral values are more useful. In view of their reputation, pyrethrin I and pyrethrin II, esters of secondary alcohols, have unexpectedly high intravenous toxicities (60) (Figure 18). In contrast, bioresmethrin, an ester of the same

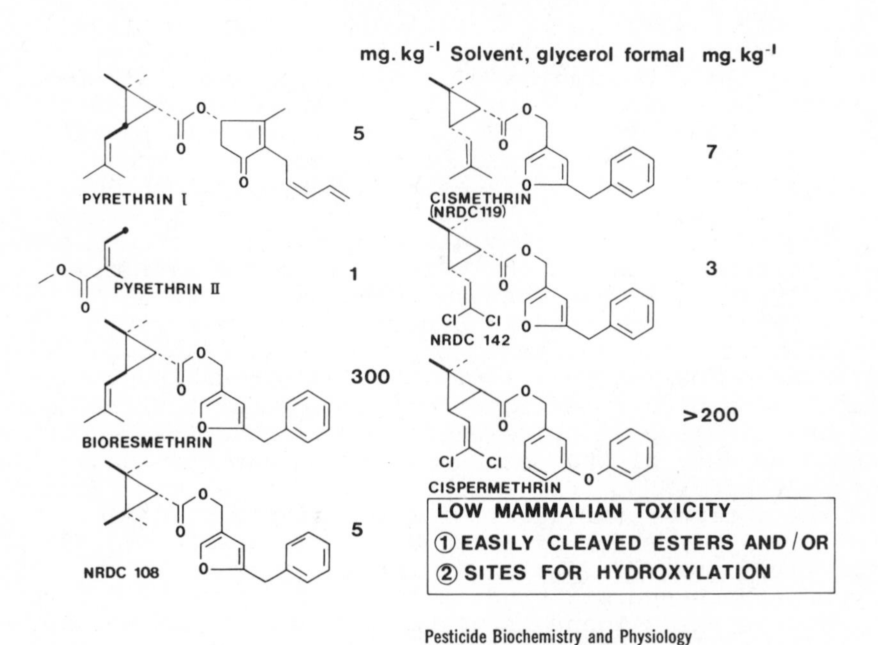

Figure 18. Intravenous toxicities to female rats (60)

trans substituted cyclopropane acid but with a primary alcohol, is some sixty times less toxic by this route (60). The same furan alcohol esterified with a cyclopropane acid having two substituents (methyl groups) cis to the carboxyl linkage (NRDC 108) (60) again has high intravenous toxicity, probably due to steric hindrance of ester cleavage. The toxicity of cismethrin (NRDC 119) (100), with isobutenyl and methyl groups cis to the ester function is similar, and is somewhat increased (in NRDC 142) (101) by substituting chlorine for methyl. However, a compound derived from NRDC 142 by changing the alcoholic component to 3-phenoxybenzyl alcohol has diminished toxicity, probably because another site, at the 4'- position on the phenoxy ring, is available for oxidative detoxification. 3-Phenoxybenzyl alcohol, therefore, is a particularly favourable alcoholic component of pyrethroids, for it is not photolabile yet gives esters of low mammalian toxicity because it can be detoxified by oxidative and esteratic attack at several sites. These concepts are considered in more detail in other papers in this symposium, and elsewhere. (102,103,104,105).

Properties of Recent Pyrethroids

Insecticides of this class developed recently, as discussed above, combine potency to insects greater than that of other categories of insecticides with lower mammalian toxicity, limited persistence in soil (69,70) and field stability adequate to control insect pests of agricultural crops. The changes in each of these properties with variation of structure have so far been considered only generally and in concluding, it is appropriate to emphasise progress in realising the insecticidal potency latent in the structure of pyrethrin I by a specific example. For this purpose, Figure 19 compares the activity of pyrethrin I with that of decamethrin, the most powerful insecticide yet synthesised. To some species of insect, decamethrin

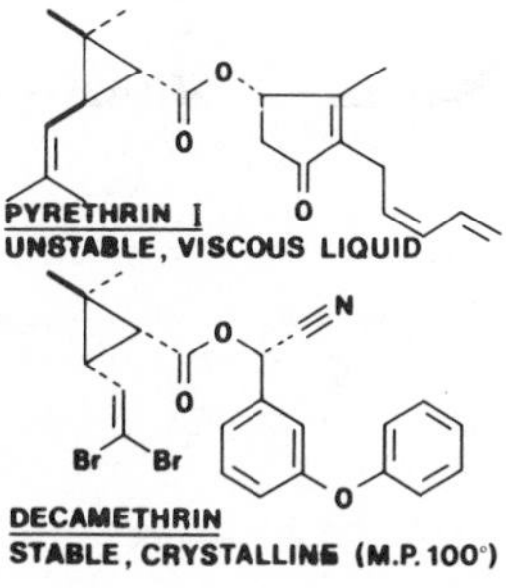

	MEDIAN LETHAL DOSES (mg.kg $^{-1}$)				
	PHAEDON COCHLEARIAE	PERIPLANETA AMERICANA	MUSCA DOMESTICA	ANOPHELES STEPHENSI	GLOSSINA AUSTENI
PYRETHRIN I UNSTABLE, VISCOUS LIQUID	0·25	0·33	16	2·4	0·37
DECAMETHRIN STABLE, CRYSTALLINE (M.P. 100°)	0·032	0·056	0·03	0·02	0·004

Figure 19. Pyrethrin I and decamethrin compared

may be several hundred-fold as active as the natural compound. Its properties, a white crystalline solid with a melting point of 100° (30), and a structure fully confirmed by X-ray analysis (106) contrast with those of pyrethrin I, which even in the purest form yet attained (95) is an unstable, viscous liquid. Figure 19 gives results of laboratory tests but field performance is equally impressive, as indicated by the results of Hadaway and Barlow (94) in an experiment to explore the potential of permethrin and decamethrin against the Tsetse fly (Table III).

Table III - Residual Toxicity to Tsetse Flies (Glossina austeni) of Insecticidal Deposits on Ivy Leaves*

Compound	Rate $g.ha^{-1}$	% Mortality of flies exposed for 1 minute to deposits (age in weeks)					
		0	1	2	3	4	6
Endosulfan	500	100	92	50	17	-	-
Dieldrin	500	100	88	54	4	-	-
Permethrin	500	100	100	100	100	100	100
	25	96	96	-	-	100	69
Decamethrin	22	100	100	100	-	100	100
	4.7	100	100	100	-	100	100
	1.0	94	-	88	-	79	-

* Data adapted from Hadaway, Barlow, Turner and Flower (94)

On a waxy leaf surface, the involatility, photostability and great insecticidal activity of these compounds result in deposits still toxic to flies at six weeks and at rates of application 1/20th, 1/100th and even 1/500th of those of endosulphan and dieldrin.

Future Prospects for Pyrethroids

These results, with compounds shown in initial studies to degrade within weeks in many soil types (69, 70) indicate the great potential of appropriate members of this group of compounds for practical applications. Important features are the many structural variations possible providing a range of useful combinations of insecticidal activity, insect species specificity, mammalian toxicity and environmental stability to match differing requirements. The rapidly developing knowledge of relationships between chemical structures and their insecticidal activities, mammalian toxicities and photostabilities suggests that continued research will disclose a number of insecticides with improved properties in this group of compounds.

Acknowledgements

I thank Mr. F. Barlow and Dr. A.B. Hadaway, and (the late) Dr. J.M. Barnes and Mr. R.D. Verschoyle for results of tests with insects and mammals, respectively and Drs. I.J. Graham-Bryce and N.F. Janes for much help in preparing this manuscript.

Literature Cited

1. Graham-Bryce, I.J., Chemy. Ind. (1976) 545.
2. Graham-Bryce, I.J., Proc. Roy. Soc. (1977) in the press.
3. Huffaker, C.B. and Croft, B.A., Environ. Health Perspec. (1976) 14, 167.
4. Metcalf, R.L. and Luckman, W.H., (editors) "Introduction to Insect Pest Management" Wiley-Interscience, New York and London, 1975.
5. Siddall, J. Environ. Health Perspec. (1976) 14, 119.
6. Wright, J.E. Environ. Health Perspec. (1976) 14, 127.
7. Cardé, R.T. Environ. Health Perspec. (1976) 14, 133.
8. Knipling, E.F. Environ. Health Perspec. (1976) 14, 145.
9. Kuhr, R.J. and Dorough, H.W., "Carbamate Insecticides: Chemistry, Biochemistry and Toxicology" CRC Press, Cleveland, Ohio, 1976.
10. Eto, M., "Organophosphorus Pesticides: Organic and Biological Chemistry" CRC Press, Cleveland, Ohio, 1974.
11. Brooks, G.T., "Chlorinated Insecticides" CRC Press, Cleveland, Ohio, 1973.
12. Metcalf, R.L. "Organic Insecticides, Their Chemistry and Mode of Action", Interscience, New York, 1955.
13. Gnadinger, C.B. "Pyrethrum Flowers" McLaughlin Gormley King Co., Minneapolis, Minnesota, 1936.
14. Gnadinger, C.B. "Pyrethrum Flowers" Supplement to Second Edition, (1936-1945). McLaughlin Gormley King Co., Minneapolis, Minnesota, 1945.

15. Nelson, R.H. "Pyrethrum Flowers" Third Edition (1945-1972) McLaughlin Gormley King Co., Minneapolis, Minnesota, 1975.
16. Casida, J.E. "Pyrethrum, the Natural Insecticide" Academic Press, New York, 1973.
17. Matsui, M. and Yamamoto, I.,in "Naturally Occurring Insecticides", (Jacobson, M. and Crosby, D.G., Eds.) Marcel Dekker, Inc., New York 1971.
18. Crombie, L. and Elliott, M. Fortschritte der Chemie Organischer Naturstoffe (1961), 19, 120.
19. Elliott, M. and Janes, N.F., in "Pyrethrum, the Natural Insecticide" (J.E. Casida, Ed.), Chap. 4. 55. Academic Press, New York, 1975.
20. Briggs, G.G., Elliott, M., Farnham, A.W. and Janes, N.F. Pestic Sci. (1974) 5, 643.
21. Briggs, G.G., Elliott, M., Farnham, A.W., Needham, P.H., Pulman, D.A. and Young, S.R. Pestic. Sci. (1976) 7, 236.
22. O'Brien, R.D. "Insecticides : Action and Metabolism" Academic Press, New York and London, 1967.
23. Corbett, J.R. "The Biochemical Mode of Action of Pesticides" Academic Press, London, 1974.
24. Krueger, H.R. and Casida, J.E. J. Econ. Ent. (1957) 50, 356.
25. Metcalf, R.L. and McKelvey, J.J. Jr. "The Future for Insecticides : Needs and Prospects" John Wiley and Sons, Inc., New York, 1976.
26. Elliott, M. in "The Future for Insecticides - Needs and Prospects" (Metcalf, R.L. and McKelvey, J.J.Jr., Eds), 163, John Wiley and Sons, Inc., New York, 1976.
27. Farnham, A.W. Pestic. Sci. (1973) 4, 513.
28. Sawicki, R.M., Elliott, M., Gower, J.C., Snarey, M. and Thain, R.M. J. Sci. Food Agric. (1962) 13, 172.
29. Sawicki, R.M. and Elliott, M. J. Sci. Food Agric. (1965) 16, 85.
30. Elliott, M., Farnham, A.W., Janes, N.F., Needham, P.H. and Pulman, D.A., Nature (1974) 248, 710.
31. Elliott, M. Bull. Wld. Hlth. Org. (1971) 44, 315.
32. Elliott, M., Farnham, A.W., Janes, N.F., Needham, P.H. and Pulman, D.A. ACS Symp. Ser. (1974) 2, 80
33. Elliott, M., Janes, N.F. and Pulman, D.A., J.C.S. Perkin 1, (1974) 246.
34. Begley, M.J., Crombie, L., Simmonds, D.J. and Whiting, D.A. J.C.S. Perkin I (1975) 1230.
35. Staudinger, H. and Ruzicka, L., Helv. Chem. Acta, (1924), 7, 177.
36. Yamamoto,R., J.Tokyo Chem.Soc. (1919), 40, 126.

37. Yamamoto, R. J. Chem.Soc.Japan.(1923), 44, 311.
38. Yamamoto, R. and Sumi, M. J. Chem.Soc.Japan, (1923), 44, 1080.
39. Schechter, M.S., Green, N. and La Forge, F.B., J. Amer.Chem.Soc. (1949) 71, 3165.
40. Roark, R.C. "A Digest of Information on Allethrin", U.S. Dept. Agric. Bur. Entomology and Plant Quarantine, Agric. Res. Admin. E 846 Sept. 1952.
41. Roark, R.C. and Nelson, R.H. "A Second Digest of Information on Alletrhin and Related Compounds." U.S. Dept. Agric., Agric. Res. Service, ARS-33-12, 1955.
42. Elliott, M., J. Sci. Food Agric.(1954) 5, 505.
43. Synerholm, M.E., U.S. Patent 2,458,656 (1949).
44. Fujimoto, K., Itaya, N., Okuno, Y., Kadota, T. and Yamaguchi, T., Agric. Biol. Chem. (1973) 37, 2681.
45. Ohno, N., Fujimoto, K., Okuno, Y., Mizutani, T., Hirano, M., Itaya, N., Honda, T. and Yoshioka, H. Agric. Biol. Chem. (1974) 38, 881.
46. Ohno, N., Fujimoto, K., Okuno, Y., Mizutani, T., Hirano, M., Itaya, N., Honda, T. and Yoshioka,H. Pestic. Sci. (1976) 7, 241.
47. Gillam, A.E. and West, T.F., J. Chem. Soc.(1942) 671.
48. Elliott, M. J. Chem.Soc. (1964) 1854.
49. Elliott, M. J. Applied Chem. (1961) 11, 19.
50. Bramwell, A.F., Crombie, L., Hemesley, P., Pattenden, G., Elliott, M. and Janes, N.F., Tetrahedron (1969) 25, 1727.
51. Crombie, L., Pattenden, G. and Simmonds, D.J., Pestic. Sci. (1976) 7, 225.
52. Elliott, M. J. Chem. Soc. (1964) 888.
53. Elliott, M. J. Chem. Soc. (1965) 3097.
54. Pattenden, G. and Storer, S., J.C.S. Perkin I, (1974), 1606.
55. Elliott, M. J. Chem. Soc. (1964) 5225.
56. Rauch, F., Lhoste, J. and Birg, M.L., Mededelingen Fakulteit Landbouw Wetenschappen Gent (1972), 37, 755.
57. Elliott, M., Needham, P.H. and Potter, C., Ann. Appl. Biol. (1950) 37, 490.
58. Elliott, M., Farnham, A.W., Janes, N.F., Needham, P.H. and Pearson, B.C. Nature (1967), 213, 493.
59. Elliott, M., Janes, N.F. and Graham-Bryce, I.J., Proc. Eighth.Br.Insec.Fung.Conf. (Brighton)(1975) 373.
60. Verschoyle, R.D. and Barnes, J.M., Pestic. Biochem. Physiol. (1972) 2, 308.

61. Lhoste, J. and Rauch, F. C.R. Acad. Sci. (Paris) (1969) 268, 3218.
62. Velluz, L., Martel, J. and Nominé, G. C.R. Acad. Sci. (Paris) (1969) 268, 2199.
63. Lhoste, J., Martel, J and Rauch, F. Proceedings of the 5th British Insecticide and Fungicide Conference (1969) 554.
64. Elliott, M., Farnham, A.W., Janes, N.F., Needham, P.H. and Pulman, D.A. Nature (1973) 244, 456.
65. Burt, P.E., Elliott, M., Farnham, A.W., Janes, N.F., Needham, P.H. and Pulman, D.A. Pestic. Sci. (1974) 5, 791.
66. Elliott, M., Farnham, A.W., Janes, N.F., Needham, P.H. and Pulman, D.A. Pestic. Sci. (1975) 6, 537.
67. Elliott, M., Farnham, A.W., Janes, N.F., Needham, P.H., Pulman, D.A. and Stevenson, J.H. Nature (1973) 246, 169.
68. Chen, Y-L and Casida, J.E., J. Agric.Food Chem. (1969) 17, 208.
69. Kaufmann, D.D., Jordan, E.G., Kayser, A.J. and Clark Haynes, S., ACS Symp. Ser. (1977) this volume
70. Barlow, F., Hadaway, A.B., Flower, L.S., Grose, J.E.H. and Turner, C.R., Pestic. Sci. (1977) 8, (in the press).
71. Miyamoto, J. Environ. Health Perspec. (1976) 14,15
72. Miyakado, M., Ohno, N., Okuno, Y., Hirano, M., Fujimoto, K. and Yoshioka, H. Agric. Biol. Chem. (1975) 39, 267.
73. Elliott, M., Farnham, A.W., Janes, N.F. and Pulman, D.A. Unpublished results.
74. Elliott, M. and Janes, N.F., ACS Symp. Ser. (1977) this volume.
75. Elliott, M., Farnham, A.W., Janes, N.F., Needham, P.H. and Pulman, D.A., Pestic. Sci. (1976) 7, 499.
76. Elliott, M., Farnham, A.W., Janes, N.F., Needham, P.H. and Pulman, D.A. Pestic. Sci. (1976), 7, 492.
77. Lhoste, J. and Rauch, F., Pestic. Sci. (1976), 7, 247.
78. Brown, D.G., Bodenstein, O.F. and Norton, S.J., J. Agric. Food. Chem. (1973) 21, 767.
79. Norton, S.J., Bodenstein, O.F. and Brown, D.G. Botyu-Kagaku (1976) 41, 1.
80. Matsuo, T., Itaya, N., Mizutani, T., Ohno, N., Fujimoto, K., Okuno, Y., Yoshioka, H. Agr.Biol. Chem. (1976) 40, 247.
81. Rauch, F., Lhoste, J. and Martel, J. Pestic. Sci. (1974) 5, 651.
82. Wickham, J.C. Pestic. Sci. (1976) 7, 273.
83. Elliott, M., Janes, N.F. and Pearson, B.C. Pestic. Sci. (1971) 2, 243.

84. Elliott, M., Farnham, A.W., Janes, N.F. and Needham, P.H. Pestic. Sci. (1974) 5, 491.
85. Elliott, M., Janes, N.F., Jeffs, K.A., Needham, P.H. and Sawicki, R.M. Nature (1965) 207, 938.
86. Elliott, M., Janes, N.F. and Pearson, B.C. J. Sci. Food Agric. (1967) 18, 325.
87. Spencer, E.Y. in "The Future for Insecticides : Needs and Prospects" (Metcalf, R.L. and McKelvey, J.J.,Jr. Eds) 295, John Wiley and Sons Inc., New York, 1976.
88. Fukuto, T.R. in "The Future for Insecticides : Needs and Prospects" (Metcalf, R.L. and McKelvey, J.J.,Jr. Eds) 313, John Wiley and Sons Inc., New York, 1976.
89. Burgen, .A.S.V., Roberts, G.C.K. and Feeney, J.C. Nature (1975) 253, 753.
90. Miskus, R.P. and Andrews, T.L. J. Agric. Food Chem. (1972) 20, 313.
91. Katsuda, Y. and Yamamoto, S. Chem. Abs. (1976) 85, 138639.
92. Elliott, M., Farnham, A.W., Janes, N.F., Needham, P.H., Pulman, D.A. and Stevenson, J.H. Proc. Seventh Br. Insec. Fung. Conf. (Brighton) (1973) 721.
93. Elliott, M. Environmental Health Perspectives (1976) 14, 3.
94. Hadaway, A.B., Barlow, F., Turner, C.R. and Flower, L.S. Pestic. Sci. (1976) 7, in the press.
95. Elliott, M. and Janes, N.F. Chemy Ind. (1969) 270.
96. Ueda, K., Gaughan, L.C. and Casida, J.E. J. Agric. Food Chem. (1974) 22, 212.
97. Ruzo, L.O., Holmstead, R.L. and Casida, J.E. Tet. Lett. (1976) 35, 3045.
98. Holmstead, R.L., Casida, J.E. and Ruzo, L.O. ACS Symp. Ser. (1977) this volume.
99. Martel, J. and Colas, R. Proceedings of the Beltwide Cotton Research Conference, Las Vegas (1976) in the press.
100. White, I.N.H., Verschoyle, R.D., Moradian, M.H. and Barnes, J.M. Pestic. Biochem. Physiol. (1976) 6, 491.
101. Barnes, J.M. and Verschoyle, R.D. (personal communication).
102. Abernathy, C.O., Ueda, K., Engel, J.L., Gaughan, L.C. and Casida, J.E. Pestic. Biochem. Physiol. (1973) 3, 300.
103. Ueda, K., Gaughan, L.C. and Casida, J.E. Pestic. Biochem. Physiol. (1975) 5, 280.
104. Casida, J.E., Ueda, K., Gaughan, L.C., Jao, L.T. and Soderlund, D.M. Arch. Environ. Contam.

Toxicol. (1975/76) 3, 491.
105. Elliott, M., Janes, N.F., Pulman, D.A., Gaughan, L.C., Unai, T. and Casida, J.E. J. Agr. Food Chem. (1976) 24, 270.
106. Owen, J.D. J.C.S. Perkin I (1975) 1865.

2

Preferred Conformations of Pyrethroids

MICHAEL ELLIOTT and NORMAN F. JANES

Rothamsted Experimental Station, Harpenden, Hertfordshire, AL5 2JQ, England

The importance of molecular shape as a factor influencing insecticidal activity of pyrethroids is well-established (1,2,3); the structure-activity relationships recognised so far are best interpreted by assuming that the whole molecular interacts at a site in the insect with specific steric requirements for optimum fit (4). Because all known active pyrethroids are flexible molecules (eg. pyrethrin 1 (Figure 1) and decamethrin (Figure 6)) most structure-activity studies frequently do not give direct information about conformations adopted at the site of action. Any method of investigating the shapes which pyrethroid molecules tend to adopt therefore deserves attention as the features revealed could persist when the molecule is acting insecticidally. One such approach used with other types of biologically-active molecule is to study the preferred conformations predicted by theoretical calculations (5). With pyrethroids especially, some information on the arrangement of the molecule in the solid state is also available, because crystal structures of several pyrethroids have now been determined by X-ray analysis (6,7,8). This paper describes preliminary attempts to identify features which may be biologically significant using a simplifed approach based on these methods.

Method

Preferred conformations can be predicted qualitatively from Dreiding molecular models by visualising atomic interactions, but for quantitative prediction, calculation of the energies involved is necessary. In the present work the following function (9) consisting simply of a repulsive and an attractive term for non-bonded interaction between 2 atoms was used:-

$$E = A \exp(-Cr) - Br^{-6}$$

Values of A, B and C for various pairs of atoms have been suggested (10). Neither this function, nor more sophisticated relationships using molecular orbitals (eg. (11))assess accurately the actual energy levels involved, but, as Hoffmann suggests (11), even the simpler functions will often indicate reliably the position of minimum energy, i.e. the preferred conformation. In this approach, the coordinates of each atom of a particular conformer are measured from the Dreiding model, whence the coordinates of any desired conformer can be generated by computer, using Gibbs' rotational matrix method (12). This process repeated for a series of rotamers, calculating the overall energy in each case, gives a rotation graph for the bond studied (see Figures 2-4, 7, 9). The vertical axis on each graph is calibrated in arbitrary energy units, based on the barrier calculated for ethane, and the horizontal axis represents a full rotation about the bond. Clearly the results for any one bond will be influenced by the disposition of the rest of the molecule, so it is important to interpret correctly the origins of the energy barriers indicated by the computation. The computer is therefore programmed to calculate the total energy for each rotamer, and in addition to list the pairs of atoms responsible for the major contributing interactions.

Results and Discussion

The procedure was first applied to pyrethrin 1 (Figure 1) which has seven single bonds about which rotation is relatively free. The rotation graph for bond 2 (Figure 2) has two maxima which arise from interference between the cis-methyl group and either

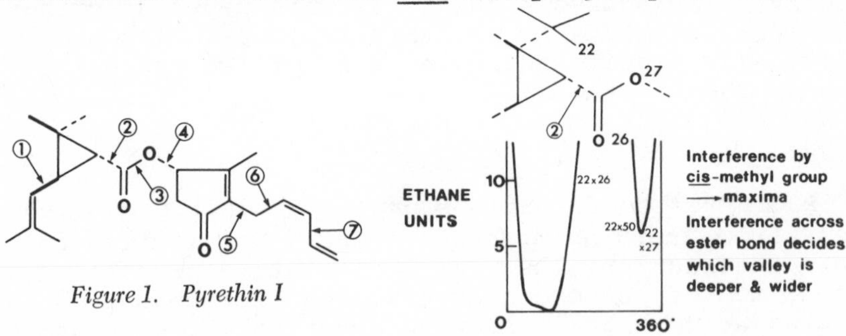

Figure 1. Pyrethin I

Figure 2. Rotation graph for bond 2

oxygen atom of the ester link. One of the corresponding valleys is filled by interactions between the cis-methyl group and the bulky alcoholic group on O-27, so the other valley is most likely to contain the preferred conformer.

For bond 3, steric considerations only (Figure 3) indicate a wide valley with the two larger groups distant from each other. However, for this bond, other considerations apply. Sutton (13) concluded that esters prefer one of two planar conformations, so that, with maximum p-π orbital overlap, some double-bond character develops in the central bond. The importance of this influence was confirmed by subsequent X-ray crystallographic studies (surveyed by Cornibert et al. (14)). In pyrethroids, with their central ester bond, the planar conformation may be very influential in determining activity as discussed earlier (1). The rotation graph (Figure 3) emphasises that probably only the transoid conformer is important in pyrethrin 1, the cisoid form being relatively hindered.

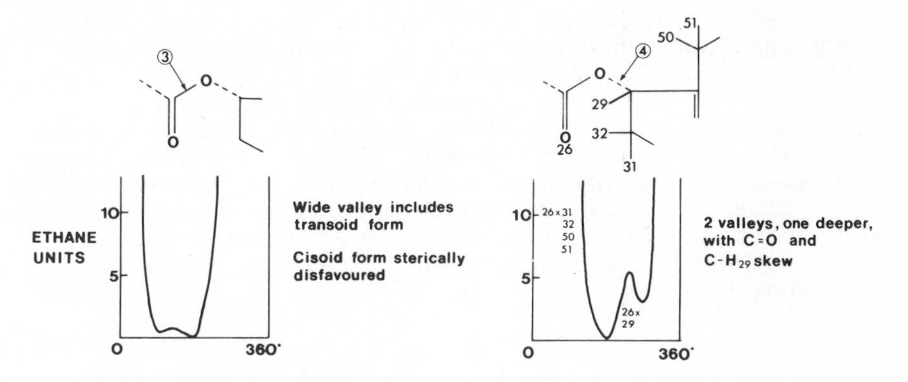

Figure 3. Rotation graph for bond 3

Figure 4. Rotation graph for bond 4

The graph for bond 4 (Figure 4) has essentially one valley, with the carbonyl group nearer to H-29 than to the larger groups on the ring. This valley is split because offsetting the carbonyl to either side of H-29 is slightly favoured energetically, and one offset position is preferred to the other.

Similar procedures were applied to the remaining single bonds in pyrethrin 1; the results are compared (Figure 5) with those actually observed (6) in the crystalline state for a closely related compound (S-bioallethrin 6-bromo-2,4-dinitrophenylhydrazone).

	1	2	3	4	5	6	7
Calculated for pyrethrin I (R=O)	94 or 214°	107	180	188	90 or 270	180	180
Found in crystalline derivative of allethrin	128°	106	180	190	99	241	—

(R = N·NH-C$_6$H$_3$(Br)(NO$_2$)NO$_2$)

Figure 5. Preferred angles in Pyrethrin I and crystals of an allethrin derivative

The two esters differ only in the length of the side-chain (allyl or pentadienyl) and in that the carbonyl group has been converted to a hydrazone in the crystalline compound. The angles for the bonds near the centre of the molecule are strikingly similar, but there are differences between the conformations calculated and observed for the side chains; possibly intermolecular forces are more significant in the crystal, and influence obtruding groups more than central ones. Also the structural differences between the compounds may affect the interactions involving the alcoholic side chain.

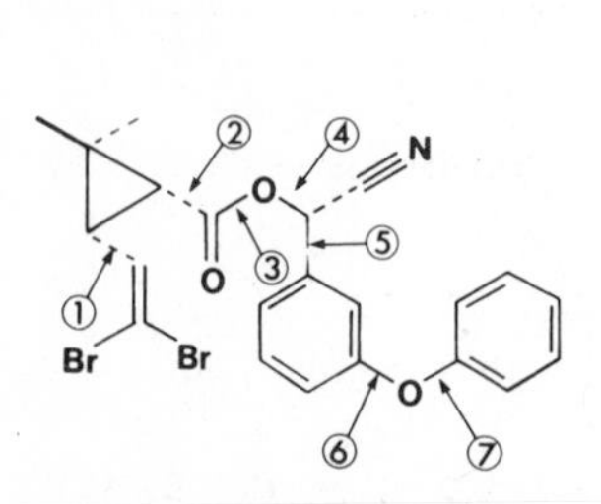

Figure 6. Decamethrin

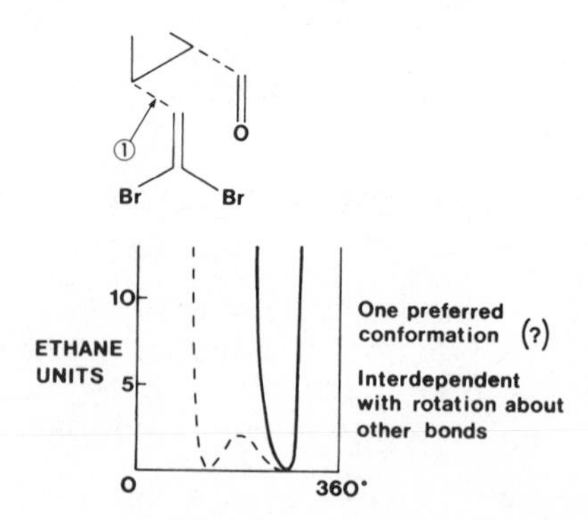

Figure 7. Rotation graph for bond 1

Decamethrin (Figure 6) is exceptionally suitable for the present purpose because it is a crystalline dibromo compound, m.p. 100°, and the structure may therefore be determined directly by X-ray analysis (7). Further, it is a considerably more powerful insecticide (15) tham most other pyrethroids, and thus probably approaches more nearly the optimum shape when it acts in the insect.

The rotation graph (Figure 7) for bond 1 in decamethrin is more complicated than in the examples above because rotations about it and about others in the molecule (particularly bond 2) are more interdependent. In one form of the molecule there is a wide valley for the side chain, but as interference increases this narrows, and there is preference for one particular conformation. The importance of this interference is uncertain but 13C magnetic resonance spectroscopy indicates that it may be significant in solution. Figure 8 shows those carbon atoms in decamethrin and its epimer at the α-carbon whose chemical shifts differ. In the fully extended molecule, the side chain carbon atoms are distant from the α-carbon and inversion there would not be expected to produce measurable shifts.

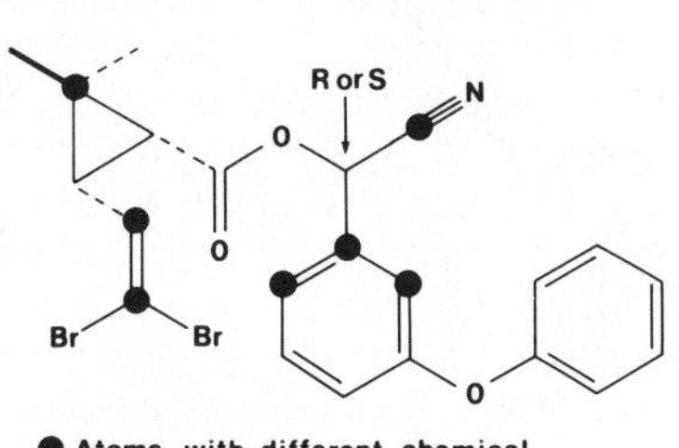

Figure 8. ^{13}C nmr spectra of decamethrin and a stereoisomer

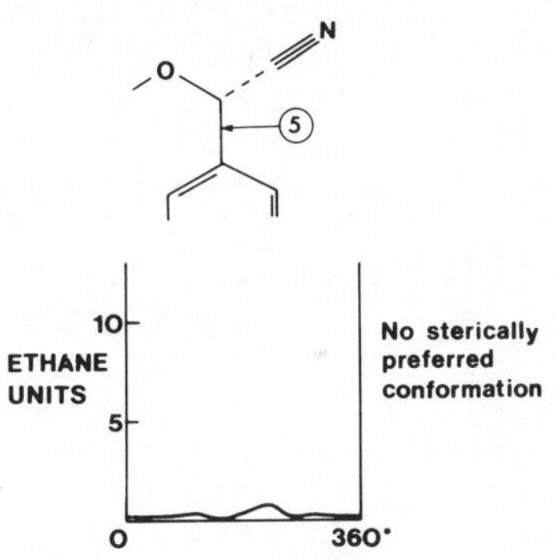

Figure 9. Rotation graph for bond 5

The differences detected suggest that the side chain spends a significant proportion of the time closer to the alcoholic part of the molecule, indicating appreciable interference such as changes the valley shape in Figure 7.

The calculations for bond 5 in decamethrin (Figure 9) indicate no preference for any particular conformer despite multiple substitution on the bond. The full results for decamethrin (Figure 10) are less definite

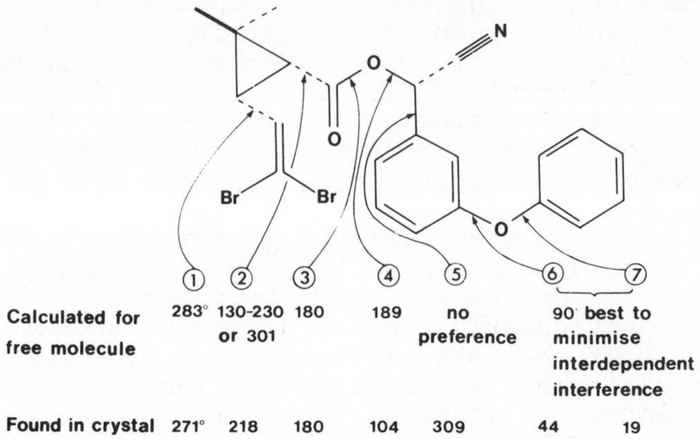

	①	②	③	④	⑤	⑥	⑦
Calculated for free molecule	283°	130-230 or 301	180	189	no preference	90 best to minimise interdependent interference	
Found in crystal	271°	218	180	104	309	44	19

Figure 10. Preferred angles in decamethrin compared with those in the crystal

than with pyrethrin 1, although the transoid form of the ester is still indicated both by calculation and by observation. Further, only bond 4 is definitely not within a valley of the rotation graph.

Thus, for some bonds, particularly those in the centre of pyrethrin 1, similarities are observed. The question therefore arises as to how significant these preferences are for insecticidal activity. Comparing the energies associated with the various perturbations to which the molecule may be subjected in the insect with the strength of conformational preference provides a partial answer. Thus, during transport the perturbations will probably be comparable to those for a molecule in solution, where thermal kinetic energy is sufficient to overcome the rotational barriers between conformers about a single bond; the nmr signals of pyrethroids in solution at ambient temperature are sharp because conformers are interconverted rapidly enough on the nmr time scale to produce a spectrum representing their weighted average. At the site of action, the molecule will experience additional forces due to binding, which may also overcome those involved in conformational preference. It is not surprising therefore that no direct relationship has been discerned. However, one concept of biological action suggests that as the molecule reaches and begins to interfere with the target, there is a crucial period during which it (or parts of it in succession (16))must achieve the correct orientation for fit; the probability of binding then depends to some extent on conformational preference. Expressed otherwise, if the optimum conformation is less easily adopted, perhaps through steric hindrance within the molecule, insecticidal activity may be diminished.

The great insecticidal activity of decamethrin may be related to lack of strongly preferred conformations about some of the important bonds, so that hindrance to the molecule adopting the optimum shape within the crucial period at the action site is minimised. Significantly, Owen showed (8) that other crystalline cis-dihalovinyl compounds, also powerful insecticides, adopt conformations about bond 1 (Figure 10) different from that in decamethrin. The insecticidal data in Figure 11 are also consistent with this suggestion. The 3-substituent can be changed quite drastically without losing insecticidal activity (17, 18); for example, a methyl group (in A) can be removed (B) and replaced elsewhere (C), except on bond 1, where it destroys activity (D & E). Calculations show that steric hindrance and conformational preference are much stronger in these last two compounds (D & E). The preferred conformation may be less appropriate, and therefore the optimum shape for fit much less easily achieved.

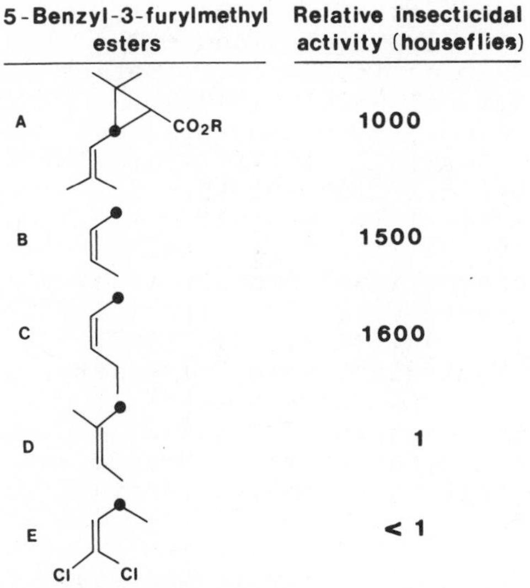

5-Benzyl-3-furylmethyl esters	Relative insecticidal activity (houseflies)
A (CO_2R)	1000
B	1500
C	1600
D	1
E (Cl, Cl)	< 1

Figure 11. A possible correlation between insecticidal activity and disturbance of conformation

Acknowledgements

We thank Dr. J.D. Owen, of the Department of Molecular Structures, Rothamsted Experimental Station, for many fruitful discussions relevant to this work and Dr. I.J. Graham-Bryce for constructive comments and continued support.

Abstract

Preferred conformations for two pyrethroids are calculated from simple models based solely on non-bonded atomic interactions. The conformations correspond closely with those found in the crystalline state for some bonds, but differ for others, particularly those near the extremities of the molecule. Despite rapid rotation about single bonds in solution, the shpae of the preferred conformation may be significant for insecticidal action, influencing the probability that the pyrethroid will adopt the appropriate shape as it approaches the target site. Some relevant biological results are discussed.

Literature Cited

1. Elliott, M. Chemy Ind. (1969) 776.
2. Elliott, M. Bull Wld Hlth Org. (1971) 44, 315.
3. Elliott, M., Farnham, A.W., Janes, N.F., Needham, P.H. and Pulman, D.A. in "Mechanism of Pesticide Action", A.C.S. Symposium Series No. 2, ed. G.K. Kohn, p. 80.
4. Elliott, M. preceding paper.
5. Kier, L.B. in "Biological Correlations - The Hansch Approach", Advances in Chemistry Series No. 114, American Chemical Society, 1972.
6. Begley, M.J., Crombie, L., Simmonds, D.J. and Whiting, D.A. J.C.S. Perkin I (1974) 1230.
7. Owen, J.D. J.C.S. Perkin I (1975) 1865.
8. Owen, J.D. J.C.S. Perkin I (1976) 1231.
9. Hill, T.L. J. Chem. Phys. (1948) 16, 399.
10. Hopfinger, A.J. "Conformational Properties of Macromolecules", Academic Press, New York, 1973.
11. Hoffmann, R. J. Chem. Phys. (1963) 39, 1397.
12. Gibbs, J.W. "Vector Analysis", Yale University Press, 1901, p. 339.
13. Sutton, L.E. in "Determination of Organic Structures by Physical Methods" (E.A. Braude and F.C. Nachod, eds.), Vol. 1, p. 405, Academic Press, New York, 1955.
14. Cornibert, J., Hien, N.V., Brisse, F. and Maxhessault, R.H. Can. J. Chem. (1974) 52, 3742.
15. Elliott, M., Farnham, A.W., Janes, N.F., Needham, P.H. and Pulman, D.A. Nature (1974) 248, 710.
16. Burgen, A.S.V., Roberts, G.C.K. and Feeney, J.C. Nature (1975) 253, 753.
17. Elliott, M., Farnham, A.W., Janes, N.F., Needham, P.H. and Pulman, D.A. Pestic. Sci. (1976) 7, 492.
18. Elliott, M., Farnham, A.W., Janes, N.F., Needham, P.H. and Pulman, D.A. Pestic. Sci. (1976) 7, 499.

3

Pyrethroid Insecticides Derived from Some Spiroalkane Cyclopropanecarboxylic Acids

R. H. DAVIS and R. J. G. SEARLE

Shell Research Ltd., Shell Biosciences Laboratory, Sittingbourne Research Centre, Sittingbourne, Kent ME9 8AG, England

The historical development of synthetic pyrethroids falls roughly into three stages. Initially there was a concentration on the structural elucidation of the natural pyrethrins, this was followed by a search for simpler alcohol components from which to form esters with the natural acids and in the last decade considerable attention has been devoted to expanding the variety of acids that can give pyrethroid esters of significant insecticidal activity on a broad spectrum of species. The success of this latter work is demonstrated by the selection of acid structures shown (Figure 1).

The structure-activity relationships derived from the work on acid components may be briefly summarised as follows:-

(a) The cyclopropane ring is not essential for activity (1).

(b) Trisubstitutedcyclopropane acids bearing unsaturated substituents give high activity; several such substituents are known and geometrical configuration can be important (2).

(c) In unsymmetrical acids activity is highly dependent on chirality (3, 4). (The only exception is 2,2-dimethylcyclopropanecarboxylic acid (5).)

(d) Geminal dimethyl groups are an essential structural requirement (3).

(e) Few tetrasubstitutedcyclopropanecarboxylic acids give active esters (6).

Although many of these observations were not available when our search for stable pyrethroids that could be used to control agricultural pests was started, it appeared that the dearth of suitable tetrasubstitutedcyclopropanecarboxylic acids afforded some scope for new synthesis. Previously the insecticidal activity of the allethronyl and pyrethronyl esters of 2,2,3,3-tetramethylcyclopropanecarboxylic acid had been described (6) and, although such esters would be too unstable for agronomic use, it was soon evident (7, 8) that several substituted benzyl esters of this acid were both active and photostable. In view of these results, the effect of replacing the geminal dimethyl groups in the tetramethylcyclopropane acid by spirofused alkane rings to give many previously unreported acids (cf. 9, 10), was systematically studied.

Synthesis of the required materials was achieved by three interrelated methods. In the first of these (Figure 2) the well-known cyclopropanation of an olefin with ethyl diazoacetate was used and in the second (Figure 3) the same olefin was reacted with chloroketene to give only one of the two possible α-chlorocyclobutanones which underwent a ready ring contraction when treated with aqueous base (11). In those cases where the appropriate olefin was not readily available, other ketene-olefin cycloadditions were used to prepare a cyclobutanone which was then halogenated prior to ring contraction to the cyclopropanecarboxylic acid. A typical sequence (Figure 4) uses methylenecyclobutane and dimethylketene as starting materials and, interestingly, in this case also the cycloaddition produces only one of the possible isomeric cyclobutanones. This is in contrast to the addition of dimethylketene to methylenecyclopropane which gives a 1 : 1 mixture of both possible products (12).

Ring contraction of α-halocyclobutanones with aqueous base offers a mechanistically interesting and useful method of preparation of cyclopropanecarboxylic acids which avoids the use of a diazoester. Although superficially similar to the well-known Favorski rearrangement of α-haloketones (13) it has been shown (11) that an unsymmetrical mechanism of the semibenzilic rearrangement type is operative. In this, hydroxide ion adds to the carbonyl carbon atom followed by a concerted displacement of halide ion by the 1,2 migration of the $C_{\alpha'}$-C carbonyl bond (Figure 5). Further evidence for an unsymmetrical mechanism is afforded by the observation that 2-bromo-2,3,3,4,4-pentamethylcyclobutanone readily contracts to the corresponding pentamethylcyclopropanecarboxylic acid. The scope of the ring contraction method thus allows mono, di, tri, tetra and pentasubstituted-cyclopropanecarboxylic acids to be prepared.

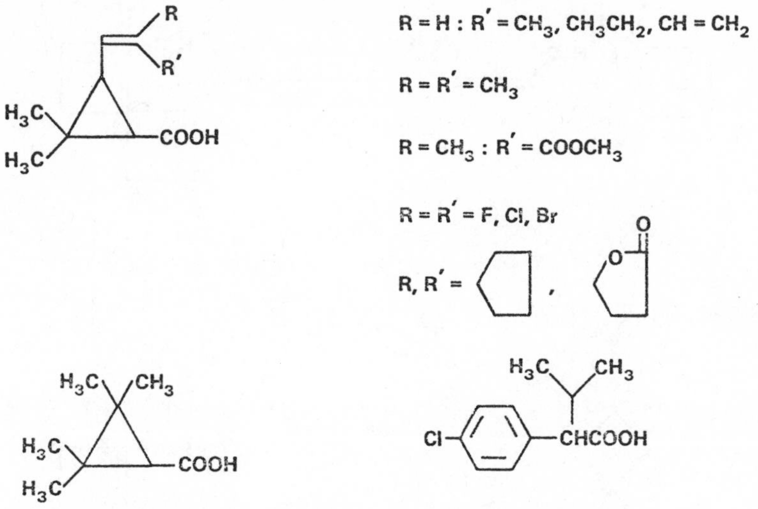
R
R'
H3C
H3C
COOH
R = H : R' = CH3, CH3CH2, CH = CH2
R = R' = CH3
R = CH3 : R' = COOCH3
R = R' = F, Cl, Br
R, R' =
O
O
H3C
CH3
H3C
H3C
COOH
H3C
CH3
Cl
CHCOOH

Figure 1. Some acids which give active pyrethroid esters

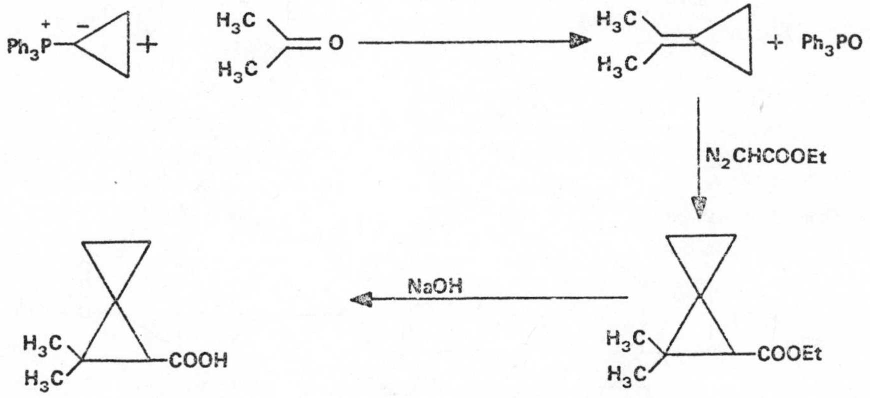
Ph3P
+
H3C
H3C
O
H3C
H3C
+ Ph3PO
N2CHCOOEt
NaOH
H3C
H3C
COOH
H3C
H3C
COOEt

Figure 2. Synthesis from diazoester–olefin addition

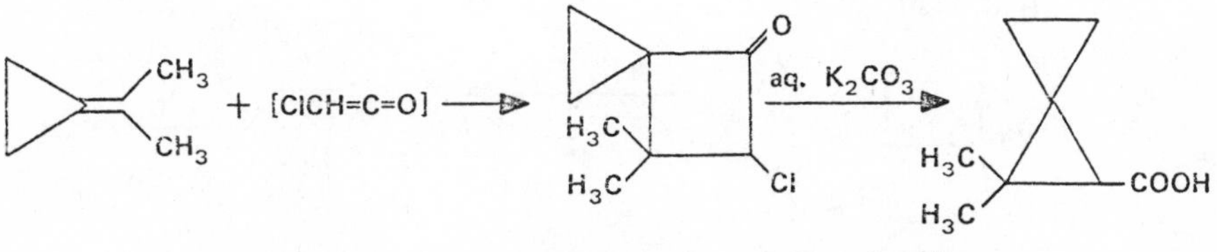
CH3
CH3
+ [ClCH=C=O]
O
H3C
H3C
Cl
aq. K2CO3
H3C
H3C
COOH

Figure 3. Synthesis from chloroketene–olefin cycloaddition

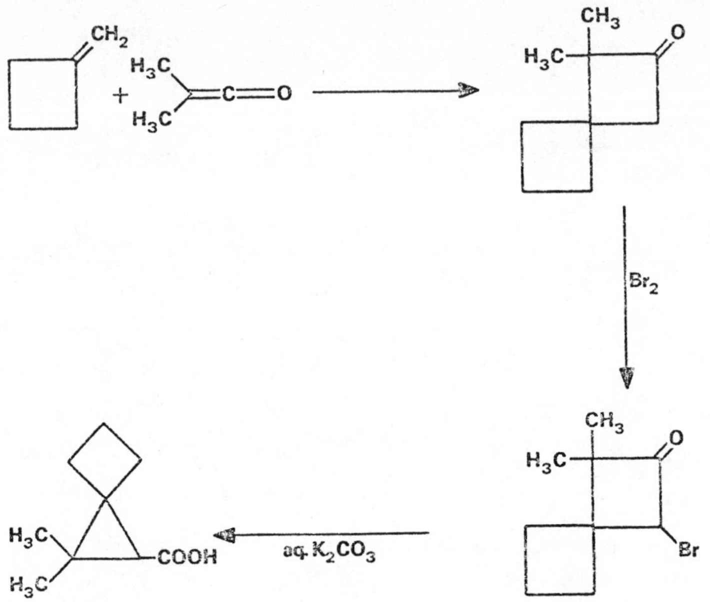

CH_2 + H_3C H_3C C=C=O → CH_3 H_3C O

Br_2

CH_3 H_3C O Br

H_3C H_3C COOH ← aq. K_2CO_3

Figure 4. Synthesis from dimethylketene–olefin cycloaddition

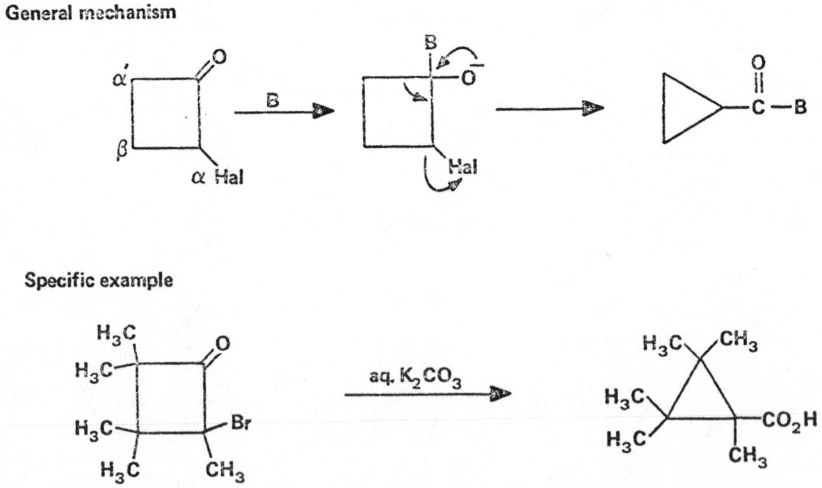

General mechanism

α′ O β α Hal → B → B O⁻ Hal → O ‖ C—B

Specific example

H_3C H_3C O H_3C Br H_3C CH_3 → aq. K_2CO_3 → H_3C CH_3 H_3C H_3C CO_2H CH_3

Figure 5. Ring contraction of α-halocyclobutanones

Compound no.	R	R_1	n	Toxicity index : Parathion = 100				Knockdown activity		
				M.d.	S.I.	T.u.	B.m.	% Concentration	KD_{50} min	KD_{90} min
I	CH_3	CH_3	2	49	600	24	45	0.025	3.8	5.2
II	CH_3	CH_3	3	44	430	18	68	0.05	4.8	6.8
III	CH_3	CH_3	4	41	170	3	76	0.1	6.4	8.9
IV	CH_3	CH_3	5	8	80	< 1		0.4	9.5	13.5
V	▷		3	38	290	< 1	14	0.025	5.2	6.5
VI	▷		4	12	330	< 1	1	0.1	5.7	10.0
VII				100	1500	24	620	0.025	2.8	3.9
	S-Bioallethrin			28	16	15	245	0.025	2.1	4.2
	Neopynamin			5	<7	4	6	0.025	2.6	5.3
	Resmethrin			55	230	5	118	0.1	4.6	7.6

T.u. = *Tetranychus urticae*

Figure 6. Comparative activities of some spiroalkane cyclopropanecarboxylates

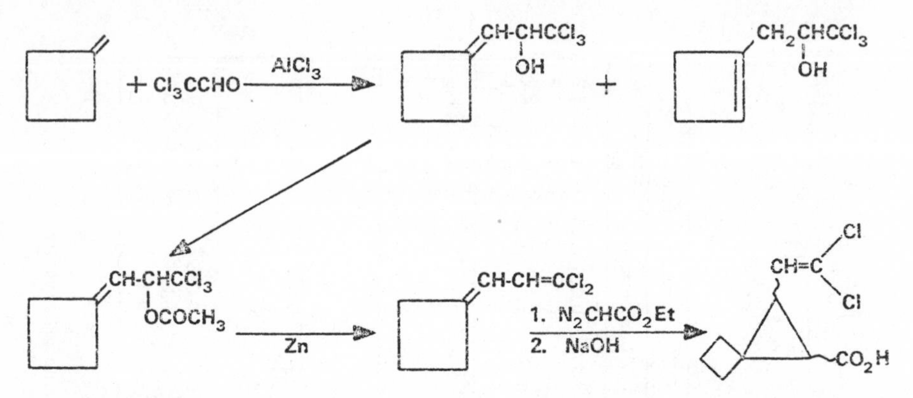

Figure 7. Synthesis route to a spiro analogue of dihalovinylcyclopropanecarboxylates

(1 : 1 cis - trans)

n	Toxicity index : Parathion = 100				Knockdown activity		
	M.d.	S.I.	T.u.	B.m.	% Concentration	KD_{50} min	KD_{90} min
3	32	530	<1	23	0.1	6.8	8.9
5	<1	<1	<1	<1	0.4	>10.0	>10.0

Figure 8. Activity of two spiroalkanedichlorovinylcyclopropanecarboxylates

Esterification of the acids prepared by the above routes with a variety of alcohols gave novel materials which have been tested on a range of insect and acarine species (Figure 6).

Several compounds, particularly those derived from α-cyano-3-phenoxybenzyl alcohol, have shown good activity on houseflies (M.d.), the cotton leafworm (S.l.) and cattle ticks (B.m.) in comparison with existing commercial pyrethroid insecticides such as S-bioallethrin and resmethrin. The knockdown effect on houseflies is also comparable with commercial standards, particularly with compound (I), and is a significant feature of many of the spiro acid esters studied.

Although in comparison with the parent tetramethyl compound (VII) the analogous spiro compounds were less toxic to insects, it was nevertheless significant, from a structure-activity viewpoint, that these compounds, some of which lack the 'essential' geminal dimethyl groups, still retained toxicity. It was thus of scientific interest to synthesise other acid types where geminal dimethyl groups in known active structures were replaced by spiroalkane rings. Thus far, some analogues of the dihalovinylcyclopropanecarboxylic acid type (14, 15) have been prepared (Figure 7). In the example shown, condensation of chloral with methylenecyclobutane gave an equal mixture of internal and external olefins and the latter was separated by column chromatography and ultimately converted into a 1 : 1 mixture of *cis* and *trans* acids. These were distinguishable by the vinylic proton which appeared in the n.m.r. specturm at δ5.95 ppm (d, J8Hz) for the *cis* acid and at δ5.33 ppm (d, J8Hz) for the *trans* compound. Esterification of two of the spirodihalovinyl acids thus prepared gave materials, the insecticidal activity of which (Figure 8) tends to reinforce previous experience with spiro compounds and may be summarised as:-

(a) Replacement of geminal dimethyl groups by a spirofused ring, within certain limits, leads to reasonable insecticidal activity on several species.

(b) Knockdown activity on houseflies is excellent and is not much affected by this modification.

(c) Limiting the size of the spirofused ring to five carbon atoms or less leads to greatest activity.

The latter observation will have to be taken into account in any comprehensive rationalisation of structure-activity relationships in pyrethroids and is reminiscent of the 'axial group' size requirement observed for activity in DDT and its analogues (16).

Acknowledgements

We thank Dr. C.B.C. Boyce and Dr. J.H. Davies for many useful discussions, Dr. J.S. Badmin for help in preparing the biological data and Shell Research Ltd., for permission to publish this paper.

Abstract

Several cyclopropanecarboxylic acids containing spirofused cycloalkane rings have been synthesised and their esters with α-cyano-3-phenoxybenzyl alcohol have been examined for insecticidal activity. Some of these compounds show moderate activity on caterpillars and rapid knockdown of houseflies. Existing structure-activity relationships in pyrethroids are reviewed in the light of these results.

Literature Cited

1. Ohno, N., Fujimoto, K., Okuno, Y., Mizutani, T., Hirano, M., Itaya, N., Honda, T., and Yoshioka, H., Agr. Biol. Chem., (1974) 38, 881.
2. Elliott, M., Farnham, A.W., Janes, N.F., Needham, P.H., Pulman, D.A., Nature, (1973) 244, 456.
3. Elliott, M., Farnham, A.W., Janes, N.F., Needham, P.H., Pulman, D.A., ACS Symposium Series No.2, (1974) 1, 80.
4. Miyakado, M., Ohno, H., Okuno, Y., Hirano, M., Fujimoto, K., Yoshioka, H., Agr. Biol. Chem., (1975) 39, 267.
5. Barlow, F., Elliott, M., Farnham, A.W., Hadaway, A.B., Janes, N.F., Needham, P.H., and Wickham, J.C., Pesticide Science, (1971) 2, 115.
6. Matsui, M., and Kitahara, T., Agr. Biol. Chem., (1967) 31, 1143.
7. Ger. Patent 2,231,312 to Sumitomo.
8. Searle, R.J.G., and Woodall, R.E., unpublished results.
9. U.S. Patent 3,823,177 to Procter Gamble.
10. Elliott, M., Bull. Wld. Hlth. Org., (1971) 44, 315.
11. Conia, J.M., and Salaun, J.R., Accounts of Chemical Research, (1972) 5, 33.
12. Isaacs, N.S., and Stanbury, P., J. Chem. Soc., Perkin II (1973), 166.
13. Kende, A.S., Organic Reactions, (1961) 11, 261.
14. Farkas, J., Kouriim, P., and Sorm, F., Chem. Listy, (1958) 52, 688.
15. Elliott, M., Farnham, A.W., Janes, N.F., Needham, P.H., and Pulman, D.A., Pesticide Science, (1975) 6, 537.
16. Holan, G., Bull. Wld. Hlth. Org., (1971) 44, 355.

4

Recent Progress in Syntheses of the New and Most Potent Pyrethroids

NOBUSHIGE ITAYA, TAKASHI MATSUO, NOBUO OHNO, TOSHIO MIZUTANI, FUMIO FUJITA, and HIROSUKE YOSHIOKA

Synthesis Laboratory, Pesticide Division, Institute for Biological Science, Sumitomo Chemical Co. Ltd., 4-2-1 Takatsukasa, Takarazuka, Hyogo 665, Japan

There have been disclosed a number of potent and photo-stable pyrethroids in which 3-phenoxybenzyl group or its α-cyano derivative are incorporated as alcohol moieties "Figure 1".

3-Phenoxybenzyl alcohol and its α-cyano derivative had originally been discovered to be useful as esters of chrysanthemic acid, i.e. S-2539 or Phenothrin (1,2), and S-2703 (3,4), the potent pyrethroids. Moreover, the former pyrethroid has been known to be more resistant to photo-irradiation than the pyrethroids with other alcohol moieties such as 5-benzyl-3-furylmethyl alcohol (5). But, the photo-stabilities of these pyrethroids had been assessed to be still insufficient under agricultural field conditions

On the other hand, the acid moieties recently developed, i.e. 2,2-dimethyl-3-(2,2-dichlorovinyl)-cyclopropanecarboxylic acid (6,7) and α-(4-chlorophenyl)-isovaleric acid (8,9) have the double bonds stabilized by the two electron-withdrawing chlorine atoms or by the aromatization forming the chloro-substituted benzene ring. 2,2,3,3-Tetramethylcyclopropanecarboxylic acid has apparently no double bond. Therefore, the esters of these acids are much more stable to photo-irradiation than the esters of chrysanthemic acid.

Among these new pyrethroids, S-3206 or Fenpropanate (3,4), NRDC-143 or Permethrin (10), NRDC-149 or Cypermethrin (11) and S-5602 or Fenvalerate (8,9) are being assessed to be most promising insecticides for agricultural use owing to their extraordinarily high potency and sufficient field persistency.

Meanwhile, there have been a variety of synthesis studies seeking for the best chemical processes for production of the respective new pyrethroids. In this connection, we now briefly review a number of known synthetic pathways and report new routes leading to the important synthetic pyrethroids and some of their essential intermediates.

Method of esterification

Figure 2 first represents a conventional method to prepare

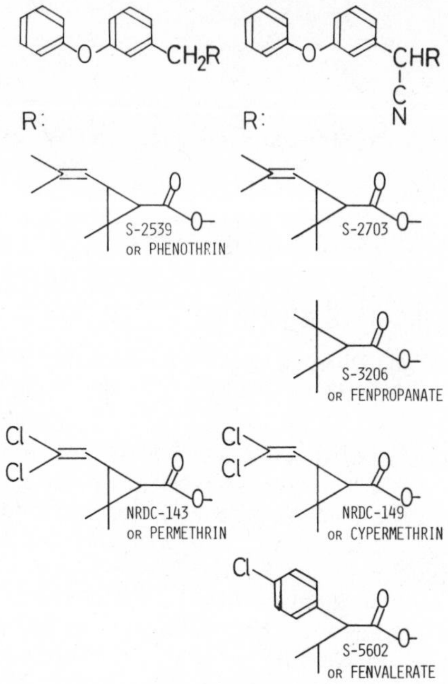

CH2R
CHR
C
N
R:
R:
S-2539
OR PHENOTHRIN
S-2703
S-3206
OR FENPROPANATE
Cl
Cl
NRDC-143
OR PERMETHRIN
NRDC-149
OR CYPERMETHRIN
Cl
S-5602
OR FENVALERATE

Figure 1. Potent synthetic pyrethroids

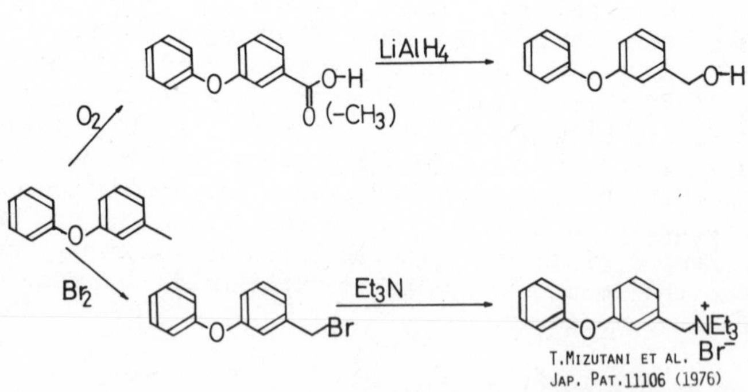

LiAlH4
O-H
(-CH3)
O2
Br2
Br
Et3N
NEt3
Br−
T. MIZUTANI ET AL. JAP. PAT. 11106 (1976)

Figure 2. Intermediates of 3-phenoxybenzyl esters

3-phenoxybenzyl alcohol, which is an intermediate of Phenothrin and Permethrin, and introduces in contrast a new method of the preparation of 3-phenoxybenzyl esters with a new intermediate.

So far, both the acid and the alcohol are primarily the requisites for the preparation of the synthetic pyrethroids. The conventional method to prepare the 3-phenoxybenzyl ester comprises the oxidation of 3-phenoxytoluene by auto-oxidation at the methyl group to form 3-phenoxybenzoic acid, the conversion of the acid into the methyl ester and the reduction of the ester with a costly hydride reagent to yield the objective alcohol, which is to be esterified with the optional acid chloride to give the final pyrethroid ester.

On the other hand, the new route comprises two characteristic steps. The first step is the preparation of the quarternary ammonium salt, 3-phenoxybenzyl triethyl ammonium bromide (12), and the second step is the condensation of the ammonium salt with the sodium salt of the optional acid (13).

The preparation of the ammonium salt is achieved by the bromination of 3-phenoxytoluene at the methyl group followed by the quarternization with triethylamine.

We initially intended to prepare pure 3-phenoxybenzyl bromide , which is to be converted into the objective ester with the sodium salt of the optional acid, but we encountered two major difficulties in this attempt. One of which was that the selectivity for the benzyl bromide was not high enough to neglect the formation of the benzal bromide and a number of by-products brominated at the aromatic nuclei. The another difficulty was lack of heat stability of the bromination mixture, that is essential for the rectification to isolate the benzyl bromide from the mixture.

These difficulties were successfully overcome by the quarternization procedure. Thus, the resulting ammonium salt can be isolated in a pure state as either crystals by filtration or an aqueous solution leaving other water-insoluble materials in an organic layer on phase separation, and the yield is quantitative on the basis of the benzyl bromide content.

In order to initiate the esterification reaction, a mixture of the aqueous ammonium salt and the sodium salt of the optional acid is boiled in a proper organic solvent to azeotropically remove water. Triethylamine is released in sequence of the esterification and a few percent of N,N-diethyl-3-phenoxybenzylamine and the ethyl ester of the adopted acid are detected in the crude product, which apparently result from the N-C bond cleavage of N-ethyl bond instead of the N-benzyl bond cleavage "Figure 3". The tertiary amine can be removed by acid-washing and the ethyl ester can be distilled off under reduced pressure. The yield of the final benzyl ester is over 90% and the purity of the benzyl ester is over 90%.

This new method is more practical than the conventional method, because the use of a costly and dangerous hydride

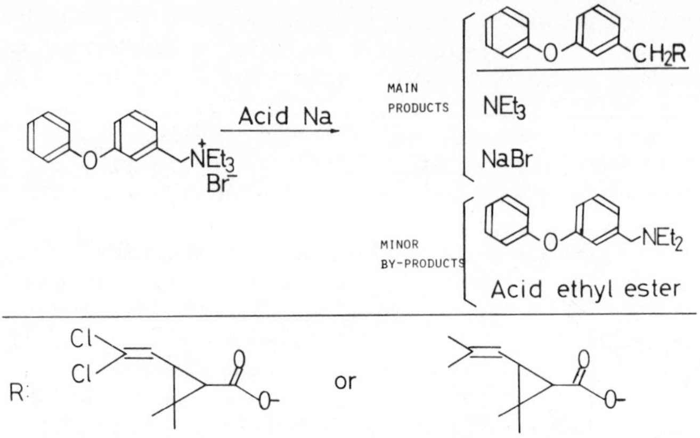

MAIN PRODUCTS

NEt_3

NaBr

Acid Na

MINOR BY-PRODUCTS

Acid ethyl ester

R: or

Figure 3. New method to prepare 3-phenoxybenzyl esters

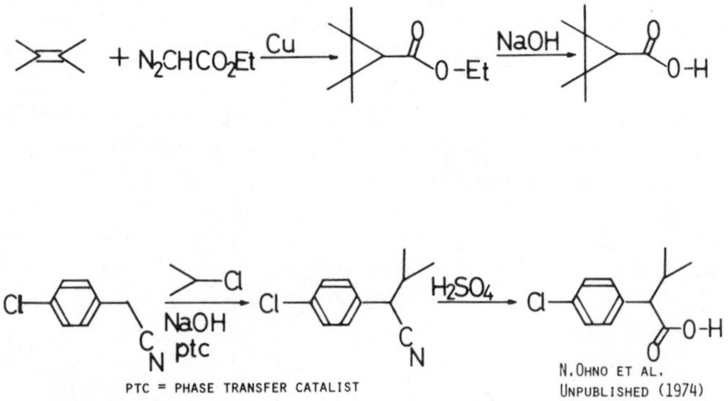

$+ N_2CHCO_2Et$ Cu O-Et NaOH O-H

H_2SO_4 NaOH ptc

PTC = PHASE TRANSFER CATALIST

N. OHNO ET AL. UNPUBLISHED (1974)

Figure 4. Syntheses of acid moieties (S-3206 and S-5602)

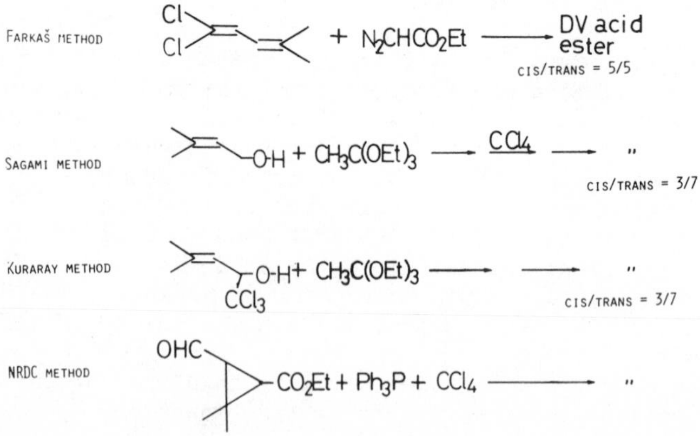

FARKAŠ METHOD $+ N_2CHCO_2Et$ → DV acid ester CIS/TRANS = 5/5

SAGAMI METHOD OH $+ CH_3C(OEt)_3$ → CCl_4 → " CIS/TRANS = 3/7

KURARAY METHOD O-H $+ CH_3C(OEt)_3$ → → " CCl_3 CIS/TRANS = 3/7

NRDC METHOD OHC $CO_2Et + Ph_3P + CCl_4$ → "

Figure 5. Syntheses of acid moieties (NRDC-143 and NRDC-149)

reagent and the preparation of the corresponding acid chloride can be avoided.

Syntheses of acid moieties

The acid moiety of S-3206, 2,2,3,3-tetramethylcyclopropanecarboxylic acid still appears to be best prepared by the carbene reaction with ethyl diazoacetate and 2,3-dimethyl-2-butene (14).

α-(4-Chlorophenyl)-isovaleric acid for S-5602 can be prepared in a quantitative yield by the hydrolysis of the corresponding nitrile, that is most simply obtained from 4-chlorophenylacetonitrile by the alkylation with isopropyl chloride and aqueous sodium hydroxide, where a phase transfer catalyst is essential to conduct the reaction (15) "Figure 4".

In contrast, the acid moiety of NRDC-143 and -149, 2,2-dimethyl-3-(2,2-dichlorovinyl)-cyclopropanecarboxylic acid, which we call DV-acid for short, can be synthesized by a number of different pathways with particular complications and different cis/trans isomer compositions "Figure 5".

Thus, in Farkaš method (7), the safe and steady handling of ethyl diazoacetate and the preparation of 1,1-dichloro-4-methyl-1,3-pentadiene appear to be major problems. In Sagami method (16) and Kuraray method (17), ethyl orthoacetate is a common requisite, which does not seem to be available yet at an economic price. Moreover, the Claisen rearrangements are to be conducted in the early stage of the methods, where ethyl orthoacetate and the respective olefinic alcohols undergo rearrangements affording the corresponding olefinic esters in moderate yields. In NRDC method (6), the Wittig reaction may be a necessary method for the preparation of particular stereoisomers.

Figure 6 shows one of the new routes leading to DV-acid, where the starting materials are the caron aldehyde ester and chloroform (18). The mixture of the aldehyde ester and chloroform is treated with potassium hydroxide to afford the chloroform adduct of the aldehyde, which is converted into the acetate with acetic anhydride and reduced with zinc dust to yield the DV-acid ester in a 65% yield. Alternatively, the chloroform adduct of the caron aldehyde ester can be synthesized from ethyl 2,2-dimethyl-3-acetylcyclopropanecarboxylate (19) by the chlorination at the acetyl methyl group. In this pathway, the Wittig reaction can be replaced by the treatments with chloroform, acetic anhydride and zinc dust.

The first step of the another route to the DV-acid is the synthesis of 3,3-dimethyl-4-(2,2-dichlorovinyl)-butanolide (20), which is obtained in a 40% yield by the condensation of 1,1,1-trichloro-2-hydroxy-4-methyl-3- or -4-pentene (21) and vinylidene chloride in the presence of a 90% sulfuric acid at a temperature around -10°C "Figure 7". The subsequent cleavage of the lactone ring with methanol and hydrogen chloride yields methyl 3,3-dimethyl-4,6,6-trichloro-5-hexenoate, which is the same intermediate as that of Kuraray method and is converted into the DV-acid ester in a high yield. Figure 8 shows an assumed mechanism of the lactone formation, in which the mixture of the trichloro alcohols (Compound VI) is converted into the more stable single isomer (Compound VI') and vinylidene chloride behaves like ethyl orthoacetate to cause the Claisen type rearrangement. In this pathway, vinylidene chloride can be utilized in place of ethyl orthoacetate.

In the light of continuing route scouting to the DV-acid, we now introduce another new pathway starting from the known enone, 4,4-dimethyl-5-hexen-2-one (22), which is obtained by either the rearrangement of the prenyl enol ether of ethyl acetoacetate followed by the hydrolysis and decarboxylation of the resulting keto ester with sodium hydroxide, or the Grignard reaction of mesityl oxide with vinyl magnesium chloride in the presence of cuprous cation as a catalyst (23) "Figure 9".

The addition of carbon tetrachloride to the enone (Compoud XII) was achieved in the presence of a 'redox catalyst' such as Cu_2Cl_2 in ethanolamine to give 4,4-dimethyl-5,7,7,7-tetrachloroheptan-2-one in more than 80% yield.

When the tetrachloro ketone (Compound XIV) was treated with aqueous methanolic sodium hydroxide at a temperature around 0°C, 2,2-dimethyl-3-(2,2,2-trichloroethyl)-cyclopropyl methyl ketone was quantitatively obtained in a 9 to 1 cis/trans ratio "Figure 10". The ring closure majorly affording the cis isomer may be explained by the assumption that the enolate anion and the trichloromethyl group behave as a bidentate ligand for the sodium cation, in analogy with the referred example where the cyclopropanedicarboxylate of cis configuration is obtained (24).

The cis cyclopropyl methyl ketone (Compound XV) has been disclosed to be the key intermediate for the DV-acid of an optional cis/trans isomer ratio by choosing two different sequences (25,26) "Figure 11".

Thus, the cis rich DV-acid was obtained through the following steps, i.e. the oxidation of the cis cyclopropyl methyl ketone with sodium hypochlorite at a temperature around 5°C afforded the corresponding trichloro acid (Compound XVI) and the elimination of hydrogen chloride from the trichloroethyl group of the acid with aqueous methanolic sodium hydroxide yielded the 90% cis DV-acid.

The 90% trans DV-acid was obtained through the other steps. The cis cyclopropyl methyl ketone (Compound XV) was first treated with sodium hydroxide in boiling methanol to convert the trichloroethyl group into the dichlorovinyl group, while the cis to trans epimerization simultaneously occurred at the C-1 carbon atom yielding the trans dichlorovinyl ketone (Compound XVII). The final step was the oxidation of the methyl ketone group with sodium hypochlorite to a carboxylic acid group to afford 90% trans DV-acid.

All these steps from the tetrachloro ketone (Compound XIV)

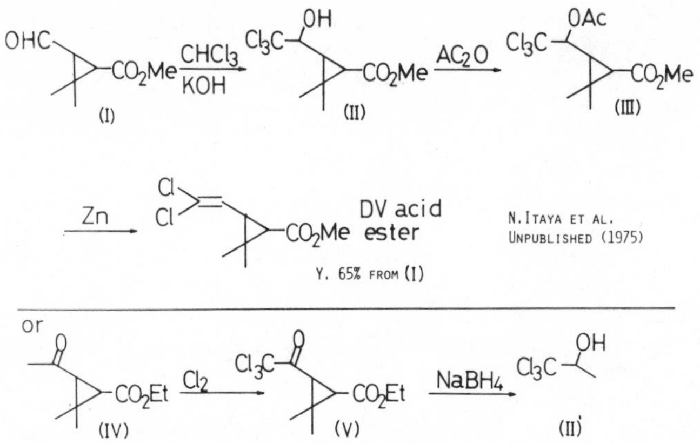

Figure 6. New method to prepare the DV acid (from caron aldehyde ester).

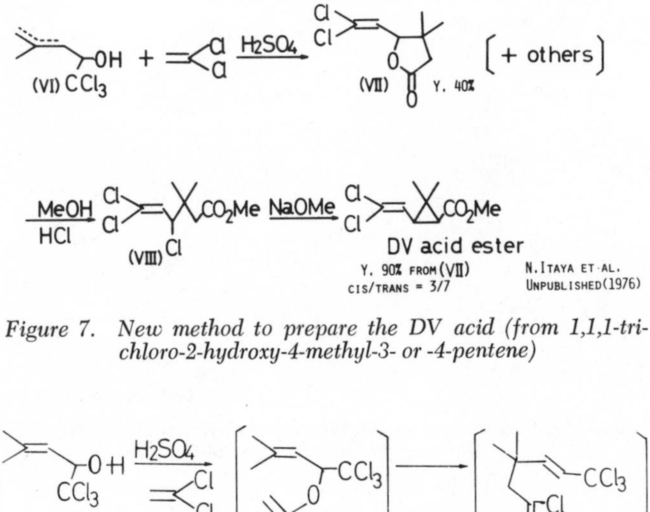

Figure 7. New method to prepare the DV acid (from 1,1,1-trichloro-2-hydroxy-4-methyl-3- or -4-pentene)

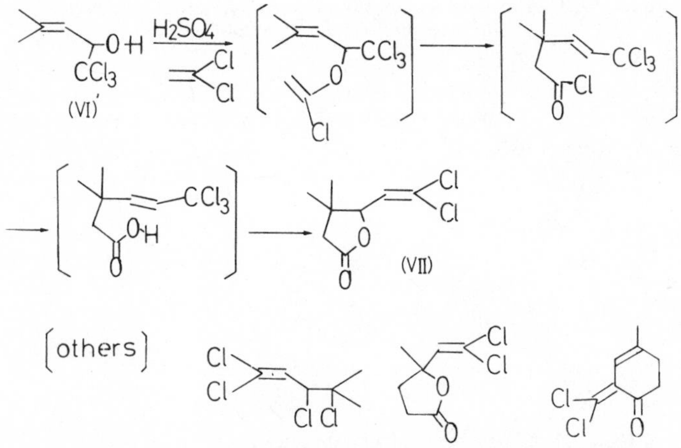

Figure 8. Assumed mechanism of the lactone formation

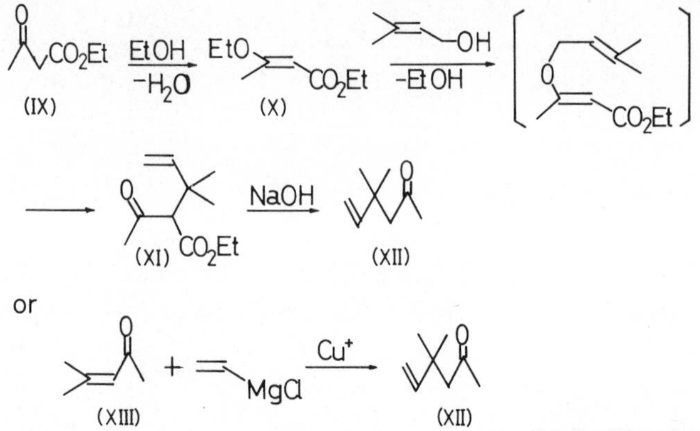

Helvetica Chimica Acta

Figure 9. Methods to prepare 4,4-dimethyl-5-hexen-2-one. The first method was presented by Brack et al. (22).

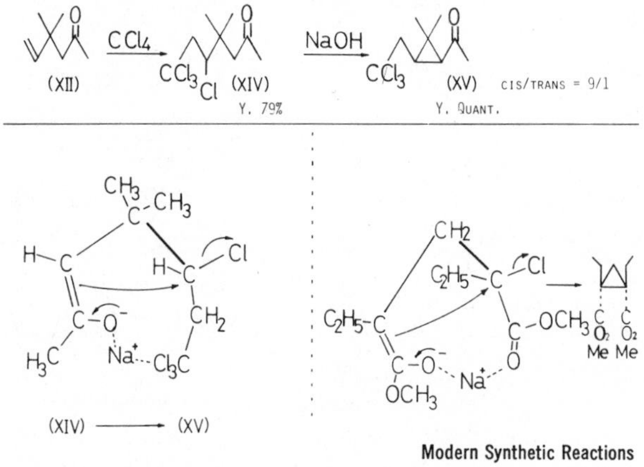

Modern Synthetic Reactions

Figure 10. Formation of cis-cyclopropylmethyl ketone and assumed mechanism (24)

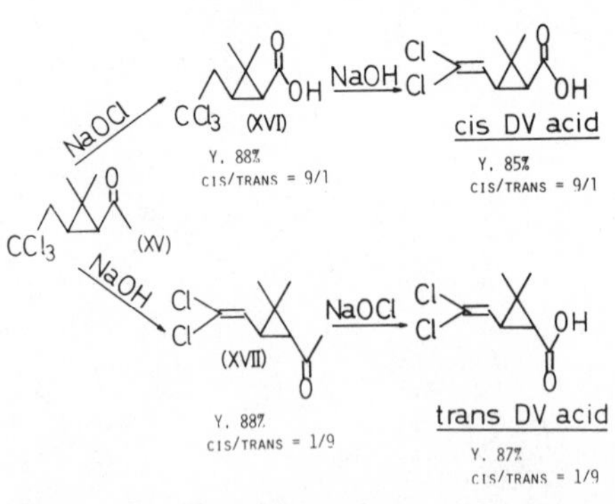

T. MATSUO ET AL. UNPUBLISHED (1975)

F. FUJITA ET AL. UNPUBLISHED (1976)

Figure 11. Formation of cis- *and* trans-*DV acid from* cis-*cyclopropylmethyl ketone* to either of the cis or the trans DV-acid may be conducted in one pot reaction. And, if the DV-acid of an optional cis/trans ratio is preferred, it may be done by a proper selection of the reaction conditions in the cyclopropane ring closure and the subsequent steps. The present method has an unique advantage over the other methods, since the pyrethroids derived from the DV-acid have different insecticidal natures, depending on the cis/trans isomer ratios.

Table I. RELATIVE TOXICITIES OF CIS AND TRANS ISOMERS

COMPD.	ISOMER	HOUSEFLY (TOPICAL APPLICATION) LD_{50}(γ/FLY)	GERMAN COCKROACH (FILM COTACT METHOD) LC_{50} (MG/M^2)
NRDC-143	CIS	0.012 (1.7)	0.88 (1.9)
	TRANS	0.020 (1.0)	1.7 (1.0)
NRDC-149	CIS	0.0032 (1.9)	0.16 (2.2)
	TRANS	0.0060 (1.0)	0.35 (1.0)

Table I represents the relative insecticidal potencies of the cis and trans NRDC-143 and -149, in which the cis isomers are nearly twice more toxic to insects than the corresponding trans isomers. Therefore, the cis pyrethroids are more preferred than the trans pyrethroids from the potency criteria and this is the first report of the selective method for the cis dominant DV-acid.

Literature Cited

1. Itaya,Nobushige; Kitamura,Shigeyoshi; Kamoshita,Katsuzo; Mizutani,Toshio; Nakai,Shinji; Kameda,Nobuyuki; Fujimoto, Keimei; Okuno,Yoshitoshi; Japan. 71 6,904
2. Fujimoto,Keimei; Okuno,Yoshitoshi; Itaya,Nobushige; Kamoshita,Katsuzo; Mizutani,Toshio; Kitamura,Shigeyoshi; Nakai,Shinji; Kameda,Nobuyuki; Japan. 71 21,473
3. Matsuo,Takashi; Itaya,Nobushige; Okuno,Yoshitoshi; Mizutani, Toshio; Ohno,Nobuo; Kitamura,Shigeyoshi; Japan. 76 5,450
4. Matsuo,Takashi; Itaya,Nobushige; Mizutani,Toshio; Ohno,Nobuo; Fujimoto,Keimei; Okuno,Yoshitoshi; Yoshioka,Hirosuke; Agr.Biol.Chem.(1976) 40, 247.

5. Fujimoto,Keimei; Itaya,Nobushige; Okuno,Yoshitoshi; Kadota, Tadaomi; Yamaguchi,Takashi; Agr.Biol.Chem.(1973) 37, 2681.
6. Elliott,Michael; Farnham,Andrew W.; Janes,Norman F.; Needham, Paul H.; Pulman,David A.; Nature,Lond.(1973) 244, 456.
7. Farkaš,Jiří; Kouřím,Pavel; Šorm,František; Chem.listy(1958) 52, 688.
8. Fujimoto,Keimei; Ohno,Nobuo; Okuno,Yoshitoshi; Mizutani, Toshio; Ohno,Isao; Hirano,Masachika; Itaya,Nobushige; Matsuo, Takashi; Japan.Kokai 74 26,425
9. Ohno,Nobuo; Fujimoto,Keimei; Okuno,Yoshitoshi; Mizutani, Toshio; Hirano,Masachika; Itaya,Nobushige; Honda,Toshiko; Yoshioka,Hirosuke; Agr.Biol.Chem.(1974) 38, 881.
10. Elliott,Michael; Farnham,Andrew W.; Janes,Norman F.; Needham, Paul H.; Pulman,David A.; Nature,Lond.(1973) 246, 169.
11. Elliott,Michael; Farnham,Andrew W.; Janes,Norman F.; Needham, Paul H.; Pulman,David A.; Pestic.Sci.(1975) 6, 537.
12. Mizutani,Toshio; Ume,Yoshitaka; Matsuo,Takashi; Japan. 76 11,106
13. Mizutani,Toshio; Ume,Yoshitaka; Matsuo,Takashi; Japan.Kokai 75 46,648
14. Matsui,Masanao; Kitahara,Takeshi; Agr.Biol.Chem.(1967) 31, 1143.
15. Ohno,Nobuo; Umemura,Takeaki; Watanabe,Tetsuhiko; Japan.Kokai 76 63,145
16. Kondo,Kiyoshi; Matsui,Kiyohide; Negishi,Akira; Takahatake, Yuriko; Japan.Kokai 76 65,734
17. Mori,Fumio; Ohmura,Yoshiaki; Nishida,Takashi; Itoi,Kazuo; Japan.Kokai 76 41,324
18. Itaya,Nobushige; Matsuo,Takashi; Magara,Osamu; (unpublished)
19. Payne,George B.; J.Org.Chem.(1967) 32, 3351.
20. Itaya,Nobushige; Fujita,Fumio; (unpublished)
21. Colonge,Jean; Perrot,André; Bull.Soc.Chim.France. 1957, 204.
22. Brack,K.; Schinz,H.; Helv.Chim.Acta.(1951) 34, 2005.
23. Von Fraunberg,Karl; Ger.Offen. 2,432,232.
24. House,Herbert O.; "Modern Synthetic Reactions. 2nd Edition" W.A.Benjamin,Inc. Menlo Park,California 1972
25. Matsuo,Takashi; Itaya,Nobushige; Magara,Osamu; (unpublished)
26. Fujita,Fumio; Itaya,Nobushige; Matsuo,Takashi; (unpublished)

5

Insecticidally Active Synthetic Pyrethroid Esters Containing a 3-(2,2-Dichlorovinyloxy)benzyl Fragment

PHILIP D. BENTLEY and NAZIM PUNJA

ICI Plant Protection Division, Jealott's Hill, Berkshire, England

Progress of synthetic pyrethroids in terms of their structure - activity relations over the last few years has led to three important acid and two alcohol fragments :-

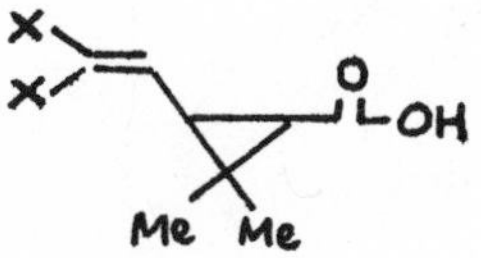

Chrysanthemic acid and its halo. analogues (x = methyl and halogen)

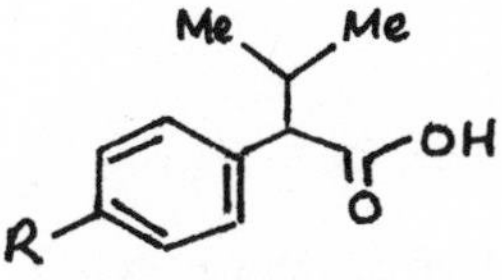

Isopropyl-4-substituted phenylacetic acid (e.g. R = alkyl, halogen)

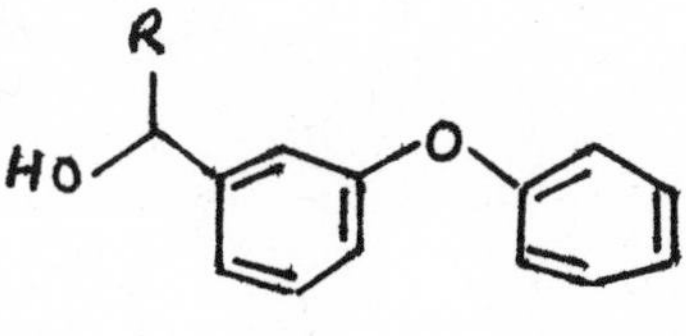

3-phenoxybenzyl and α-cyano alcohol (R = H and CN)

The combination of each of the acids with the corresponding alcohols has thus produced insecticidally active pyrethroid esters e.g. :-

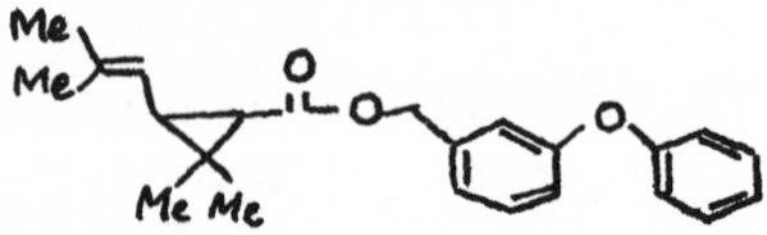

phenothrin

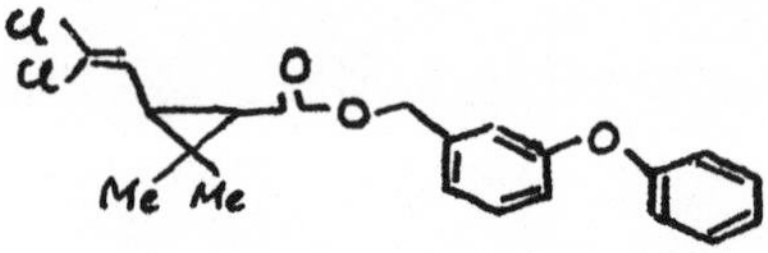

NRDC 143

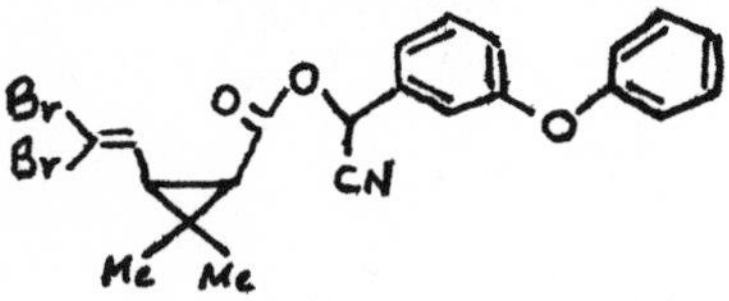

NRDC 161

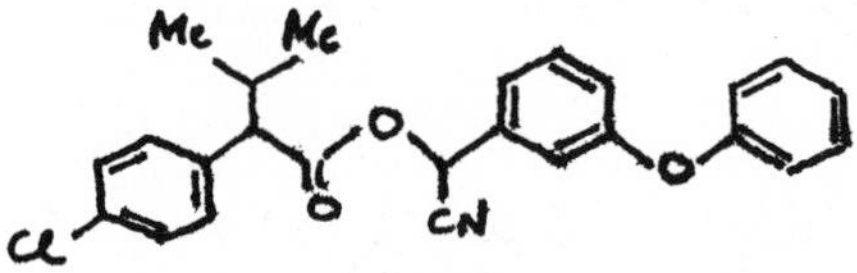

S 5602

Examination of the alcohol fragment shows that both these alcohols (and many others) comprise of a primary alcohol or cyanohydrin attached to an aromatic or hetero-aromatic ring to which is also attached in a 1,3-arrangement a freely rotating phenoxy or benzyl group.

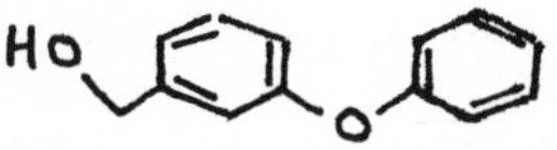

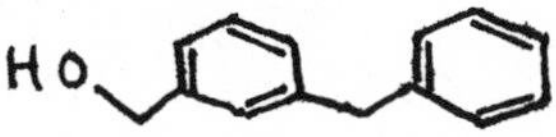

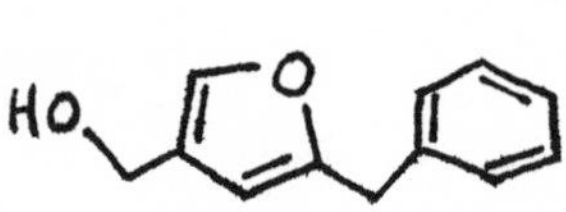

We wished to investigate this freely rotating phenoxy group and to replace it with the dichlorovinyloxy function to give, e.g. :-

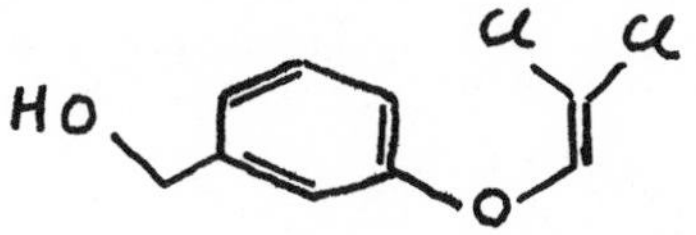

A molecular model of 3-(2,2-dichlorovinyloxy) benzyl alcohol retained the structural, spacial and rotational requirements believed to be essential for insecticidal activity.

1,2-Elimination in a trichloroethyl group to give 1,1-dichlorovinyl group has been applied, using zinc, to several systems. There are notably two examples of such a reductive elimination[1,2] :-

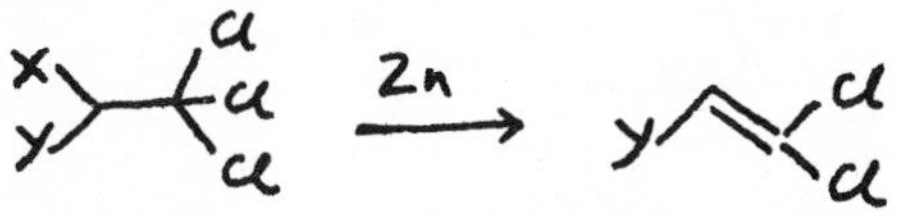

($X = Y = OAc$ or $X = SO_2Et$, $Y = OCONHMe$)

We, therefore, needed as our precursor, either the acetate or the sulphone moitey :-

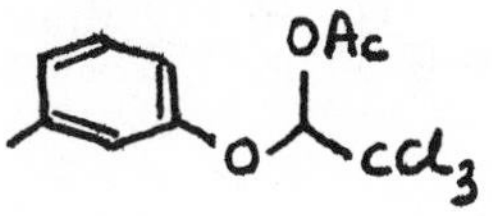

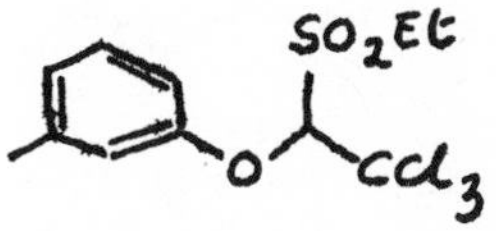

A paper by Von Hess and Moll[3] which described the reaction of substituted phenols with anhydrous chloral in the presence of acetyl chloride gave us a direct entry into this synthesis.

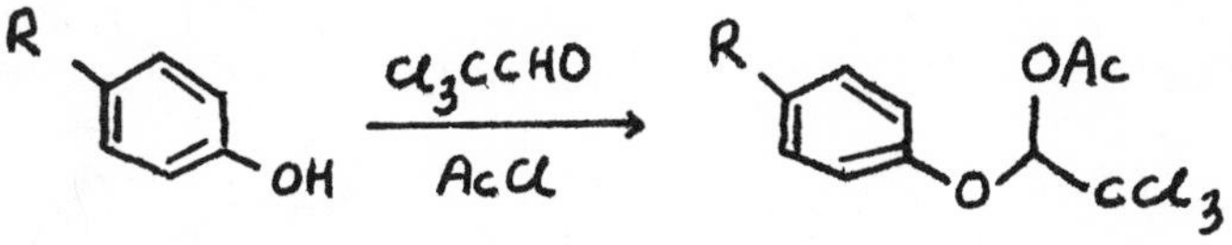

($R = H, Cl$)

Addition of 3-cresol to an ethereal solution of anhydrous chloral at room temperature, followed by triethylamine and acetyl chloride, gave 1-acetoxy-2,2,2-trichloroethyl 3-tolyl ether.

The latter was dissolved in glacial acetic acid and reacted with zinc dust at ca. 50°, to give 3-(2,2-dichlorovinyloxy) toluene.

Bromination of the 3-(2,2-dichlorovinyloxy) toluene with N-bromosuccinimide gave 3-(2,2-dichlorovinyloxy) benzyl bromide. There was no evidence of nuclear bromination.

CH_3 OH Cl_3CCHO AcCl OAc CH_3 O CCl_3 Zn Br Cl Cl N.B.S. Cl Cl CH_3 O N N N N Cl Cl OHC O $Br^{\ominus}$ $\oplus$ N N N N Cl Cl O $H_3O^{\oplus}$ HCN HO CN Cl Cl O

Reaction of the bromide with the potassium carboxylate (prepared from anhydrous potassium carbonate and the acid in acetone) of chrysanthemic acid, 3,3-dimethyl-2-(2,2-dichlorovinyl) cyclopropanecarboxylic acid and isopropyl-4-tolyl acetic acid gave the corresponding pyrethroid esters related to their insecticidally active analogues.

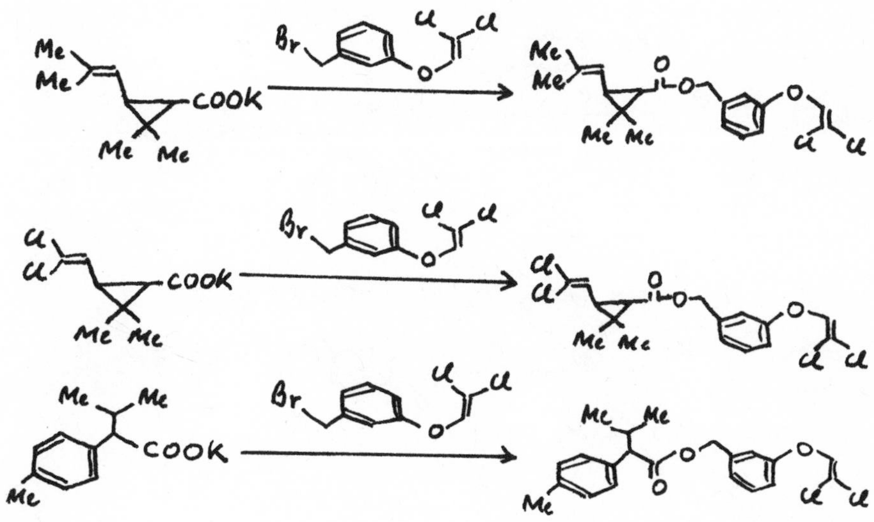

Sommelet reaction of the bromide with hexamethylene tetramine gave a crystalline quarternary ammonium salt, which upon treatment with aqueous acetic-hydrochloric acid, gave 3-(2,2-dichlorovinyloxy)-benzaldehyde.

The aldehyde upon treatment with hydrogen cyanide gave 3-(2,2-dichlorovinyloxy)-benzaldehyde cyanohydrin.

The cyanohydrin was in turn reacted with the acid chloride (prepared from the acid and thionyl chloride) of chrysanthemic acid, 3,3-dimethyl 2-(2,2-dichlorovinyl) cyclopropanecarboxylic acid and isopropyl 4-tolyl and 4-chlorophenyl acetic acid gave the corresponding pyrethroid esters related to their insecticidally active analogues.

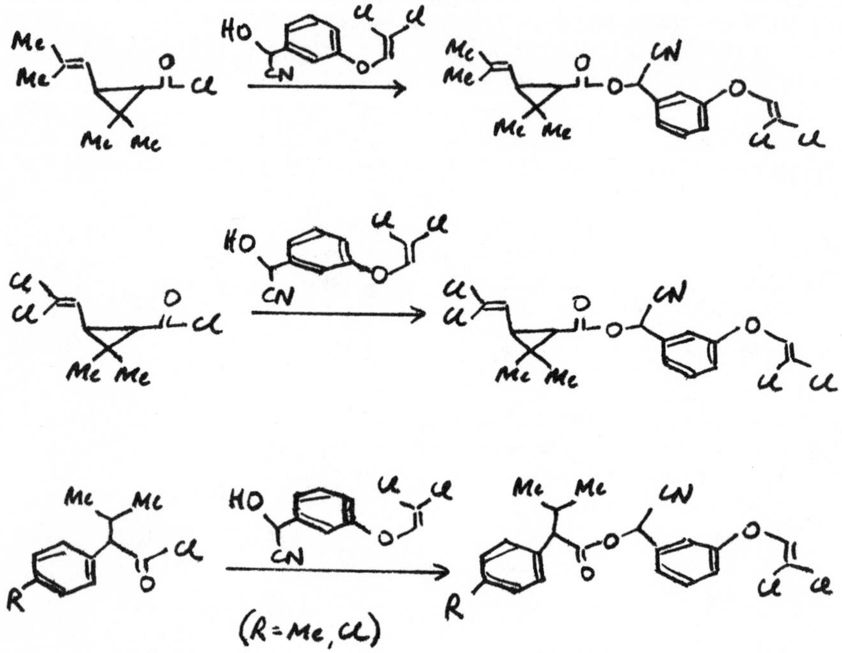

(R=Me, Cl)

By analogous procedure, the corresponding 2- and 4-(2,2-dichlorovinyloxy) benzyl esters were also prepared.

Structure-Activity Relations

The following generalisation can be made.

1. As in the NRDC series with 3-phenoxybenzyl alcohol, only the pyrethroids containing the 3-(2,2-dichlorovinyloxy) benzyl fragments are insecticidally active. The 2- and 4-substituted compounds are inactive.

2. As in the NRDC series, the 3-(2,2-dichlorovinyloxy) benzyl fragment attached to the cis -3,3-dimethyl-2-(2,2-dichlorovinyl) cyclorpropanecarboxylic acid gives more active pyrethroid esters than with the trans - acid. D-acid gives higher activity than the DL-acid.

3. As in the NRDC series, the cyanohydrin gives more active pyrethroid esters than the primary alcohol.

4. On representative test insect species (Plutella, Phaedon, Musca and Aedes), the insecticidal activity of the pyrethroid esters containing the 3-(2,2-dichlorovinyloxy) benzyl fragment are somewhat less active than their corresponding NRDC counterparts.

5. The spectrum of activity of these pyrethroid esters correspond to that of the NRDC series of pyrethroids, e.g. activity is good against Lepidoptera, Coleoptera and Diptera, fair against Homoptera (aphis) and poor against mites.

6. The rationale for structure-activity relationship thus parallels that of the NRDC pyrethroids containing the 3-phenoxybenzyl fragment. The change from phenyl to dichlorovinyl group thus only affects the degree and not the nature and spectrum of insecticidal activity.

References

1. Deodhar G. W., J. Indian Chem. Soc., (1934), 11, 83.
2. Kay I. T. and Punja N., J. Chem. Soc. C, (1968) 3011.
3. Hess Von B. and Moll R., J. Prakt Chem. (1974), 316 (2), 304.

6

Pyrethroid-Like Esters of Cycloalkane Methanols and Some Reversed-Ester Pyrethroids

MALCOLM H. BLACK

Wellcome Research Laboratories, High St., Berkhamsted, Hertfordshire, HP4 2DY, England

Two aspects of structure-activity relationships of synthetic pyrethroids are presented in this paper.

Pyrethroid-Like Esters of Cycloalkane Methanols.

A common feature in the alcoholic components of more effective pyrethroid esters (e.g. pyrethrin I, resmethrin and phenothrin) is an unsubstituted side chain supported by a planar ring containing at least one olefinic bond. However, it has been claimed (1,2) that an olefinic group can play a similar role to the cyclic nucleus; for example

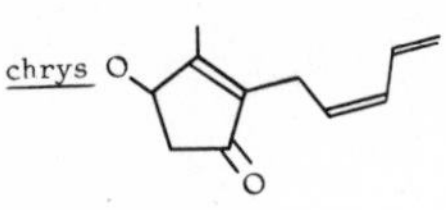

Pyrethrin I

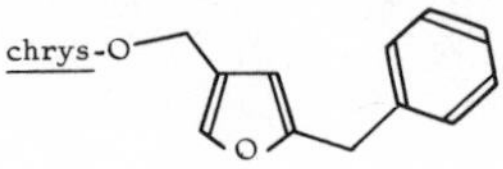

Resmethrin

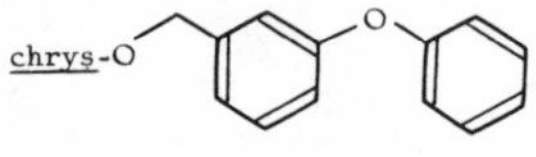

Phenothrin

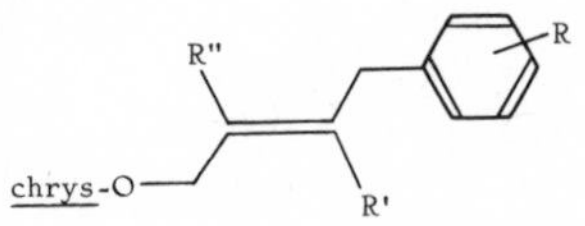

I

II

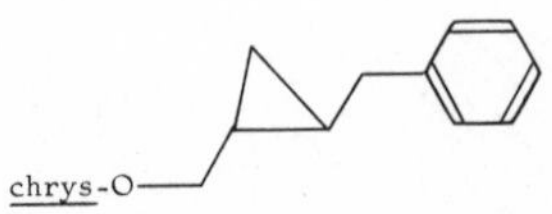

III

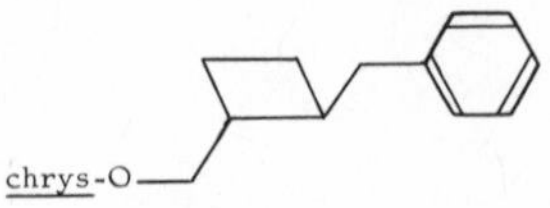

IV

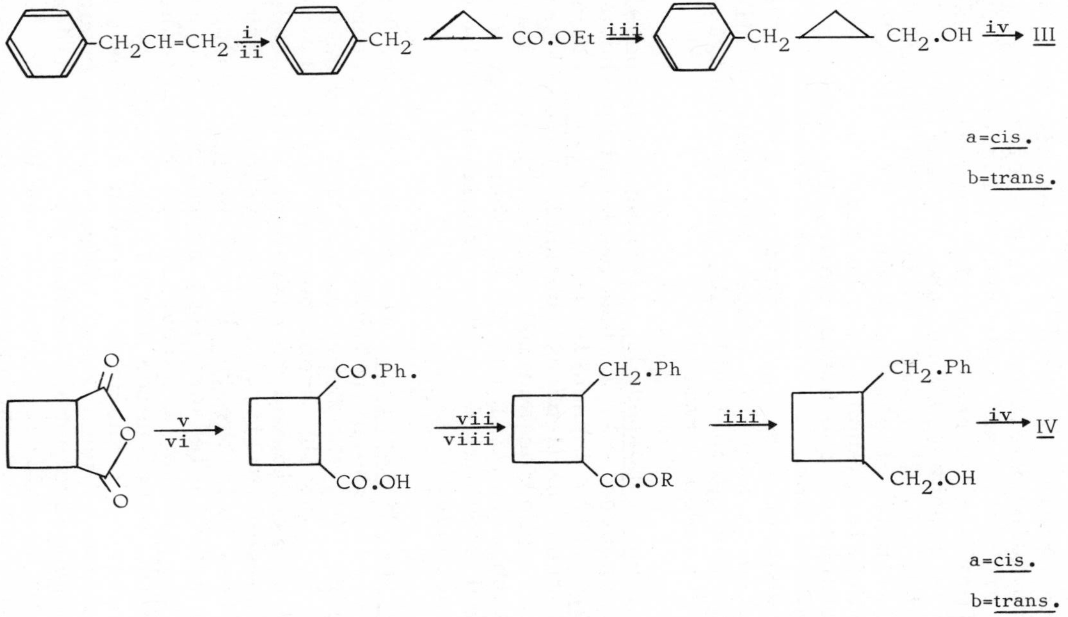

C_6H_5-CH_2CH=CH_2 →(i, ii) C_6H_5-CH_2-(cyclopropane)-CO.OEt →(iii) C_6H_5-CH_2-(cyclopropane)-CH_2.OH →(iv) III

a=cis.
b=trans.

(cyclobutane dicarboxylic anhydride) →(v, vi) (cyclobutane: CO.Ph., CO.OH) →(vii, viii) (cyclobutane: CH_2.Ph, CO.OR) →(iii) (cyclobutane: CH_2.Ph, CH_2.OH) →(iv) IV

a=cis.
b=trans.

Figure 1. Reagents: (i) *E.D.A./Cu;* (ii) *chromatographic separation of isomers;* (iii) *$LiAlH_4$;* (iv) *(+)-*trans-*chrysanthemoylchloride/pyridine;* (v) *$AlCl_3/C_6H_6$;* (vi) *AqNaOH;* (vii) *$H_2/Pd/HClO_4$ [R = H];* (viii) *$MeOH/H_2SO_4$ [R = Me]*

4-aryl-trans-2-butene-1-yl chrysanthemates (I, R=R"=H, R'=Cl) have been shown to be more effective against houseflies than allethrin (1). The corresponding cis-butenyl esters were not detectably active.

It was, therefore, of interest to ascertain whether the potency of the trans-butenyl chrysanthemates could be retained without the olefinic bond, solely by providing the alcohol with a rigid trans arrangement of benzyl and hydroxymethyl groups. An indication that such a result might be realised was suggested by a report (3) that the 4-phenylbutyl chrysanthemate, II, was twice as toxic to houseflies as the 4-phenyl-2-butenyl chrysanthemate (I, R=R'=R"=H). Although surprising, such a result might be attributed to the preference of the butyl ester, II, to exist in a transoid conformation.

As the desired structural features are provided by small saturated carbocyclic rings, the corresponding cis and trans-cyclopropyl, IIIa,b, and cyclobutyl, IVa,b, analogues were examined.

Preparation of Compounds. 4-Phenyl-2-butene-1-yl (+)-trans-chrysanthemate, I, was prepared by the method of Sota et al., (1), 4-phenyl-butyl (+)-trans-chrysanthemate (II) from the 4-phenyl-butanol(4) and (+)-trans-chrysanthemoyl chloride and the cycloalkylmethyl (+)-trans-chrysanthemates III and IV as shown (Figure 1). Analytical and spectral data were consistent with the proposed structures.

Results and Discussion. The activities for kill of houseflies, are shown in Table I. In disagreement with the report (3), the butyl ester, II, was only one quarter as potent as the trans-butenyl ester, I. This result could be attributed to different test methods and species susceptibility. Of the cycloalkylmethyl chrysanthemates, only the trans-cyclopropyl analogue, IIIb, showed detectable activity, approximately half that of the trans-butenyl ester, I, but significantly greater than the butyl ester II. If the activity of the trans-cyclopropyl analogue depended upon the rigid trans arrangement of benzyl and hydroxymethyl groups, the related trans-cyclobutyl analogue, IVb, would also be expected to be active. Since it was actually less active, the potency of the trans-cyclopropyl compound is probably associated with the nature of the cyclopropanyl ring, the hybrid orbitals of which, unlike those of cyclobutane, can provide it with some of the characteristics of an olefinic bond, apparently necessary for useful activity within this series of

Table I.

chrys.O.CH_2 – X – CH_2–⟨benzene ring⟩

Compound	—X—	Isomer	LD50 γ/♀		Relative Potency Synergised
			Alone	+P.B.(1:5)	
I	-CH=CH-	trans	0.43	0.3	40
II	$-CH_2-CH_2-$	—	c.6.0	1.1	11
IIIa	-CH—CH- (cyclopropane ring)	cis	>6.0	>2.4	<5
IIIb	-CH—CH- (cyclopropane ring)	trans	3.4	0.7	17
IVa	-CH—CH- (cyclobutane ring)	cis	>6.0	>2.4	<5
IVb	-CH—CH- (cyclobutane ring)	trans	>6.0	>2.4	<5
Bioallethrin	—	—	0.42	0.12	100

compounds.

Reversed Ester Pyrethroids.

Pyrethroid-like esters of cyclopropane methanols (5,6,7) and other related compounds (Table II) have been examined to determine the effect on activity of modifying the ester link. All were considerably less potent than the parent esters and a carbonyl group adjacent to the cyclopropane ring appears to be essential for pyrethroid-like activity. However the compounds that lack this appropriately placed carbonyl group also lack the spatial characteristics of the normal ester linkage.

Therefore, in this work, pyrethroid-like compounds (Table III, V-VIII) retaining the spatial characteristics of the normal ester linkage but lacking carbonyl groups adjacent to the cyclopropane ring, were examined. Reversal of the ester linkage provides these features and also retains a region of polarity in the linkage

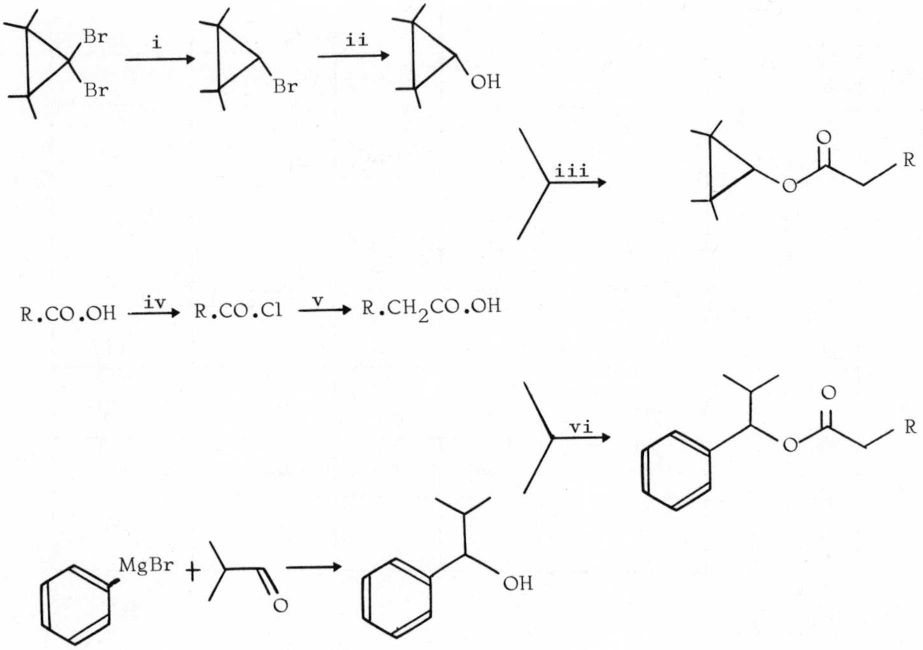

Br
Br
i
Br
ii
OH
iii
O
O
R

$$R.CO.OH \xrightarrow{iv} R.CO.Cl \xrightarrow{v} R.CH_2CO.OH$$

vi
O
O
R
MgBr +
O
OH

Figure 2. Reagents: (i) $(n\text{-}Bu)_3SnH$; (ii) Li/O_2 [−70°]; (iii) $ClCOOEt/Et_3N$; (iv) $SOCl_2/C_6H_6$; (v) CH_2N_2/Ag_2O; (vi) TsCl/pyridine [R = 1-benzyl-3-furyl and Me-phenoxyphenyl]

Table II.

COMPOUND	R.T.	COMPOUND	R.T.
tet-CO-NH-CH_2-bf	130	chr-CH_2-O-CO-bf	<1
tet-CO-CH_2-CH_2-bf	47	chr-NH-CO-O-CH_2-bf	<1
tet-CH_2-CO-O-all	<1	chr-NH-CO-O-all	<1
tet-CH.OH-CH_2-CH_2-bf	<1	chr-CH_2-O-CO-CH_2-mdp	<1
chr-CO-CH_2-all	16	chr-CO-O-all (allethrin)	100
chr-CH_2-O-all	<1	tet-CO-O-CH_2-bf (NRDC 108)	1600

R.T. = Relative toxicity (Allethrin = 100).

tet = 2,2,3,3-Tetramethylcyclopropyl.

bf = 5-Benzyl-3-furyl.

all = 2-Allyl-3-methylcyclopent-2-ene-4-yl.

chr = 2,2,-Dimethyl-3-(2-methylpropenyl)-cyclopropyl.

near to that of the normal ester.

Preparation of Reversed Esters. The compounds were synthesised as shown in Figure 2. Analytical and spectral data were consistent with the proposed structures. The α-isopropyl-phenylacetate, IX and NRDC 108 were prepared by reported procedures (7,8).

Results and Discussion. The activities, against houseflies, both alone and synergised with piperonyl butoxide are shown in Table III. All the reversed ester pyrethroids are less potent than the parent esters; however, they are significantly more active than previously reported compounds that lack a carbonyl group adjacent to the cyclopropane ring. It was significant that with both cyclopropyl and phenyl-

Table III.

No.	COMPOUND	ESTER LINK	LD50 γ/♀		F.o.S.
			ALONE	+P.B.(1:5)	
V	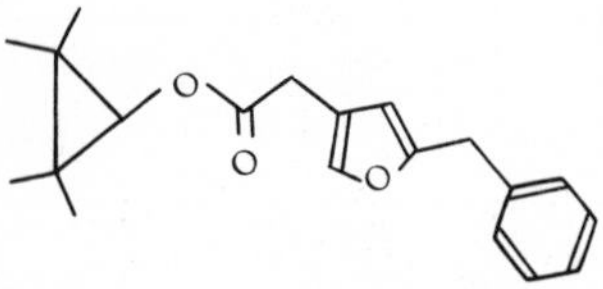	R	5.14	1.35	3.8
VI	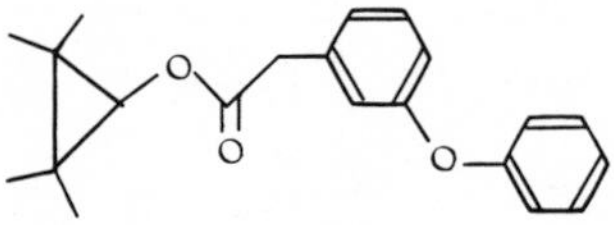	R	>6.0	>2.4	—
NRDC 108	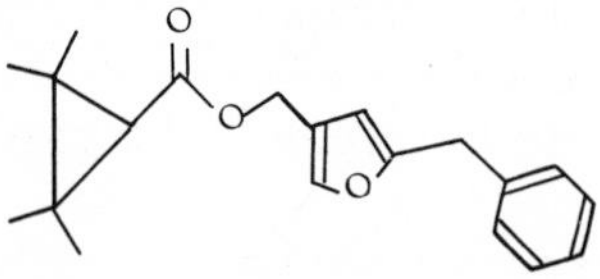	N	0.022	0.012	1.8
VII	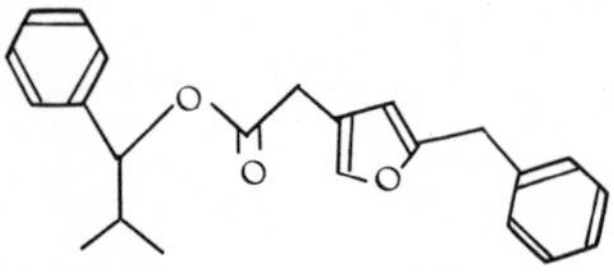	R	4.8	0.98	4.9
VIII	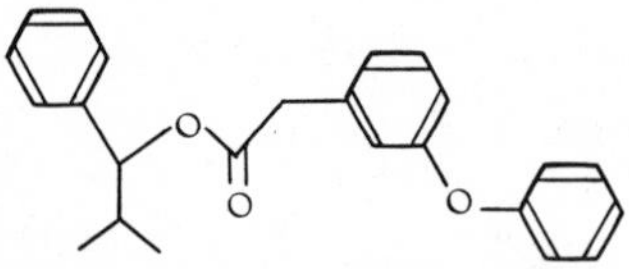	R	c.6.0	1.35	4.5
IX	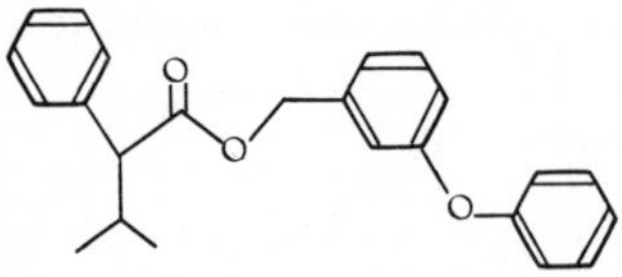	N	0.475	0.1	4.75
	BIOALLETHRIN		0.42	0.12	3.5

propyl esters, the benzyl furan moeity conferred greater activity than the phenoxyphenyl moiety. The phenylpropyl ester, VIII, has only one-tenth the activity of the parent ester, IX; the cyclopropyl ester, V, suffered an even greater reduction (c.1/100th). This suggested that the potency of reversed esters might be improved further by using other synergists.

The compounds were therefore tested with FMC 16824, a synergist claimed (9) to be highly effective for tetramethrin. As shown in Table IV, increase in activity was realised with the cyclopropyl furanacetate V. This compound was tested further with an analogous synergist FMC 11523. A similar and slightly improved result was obtained, revealing an activity approaching that of bioallethrin.

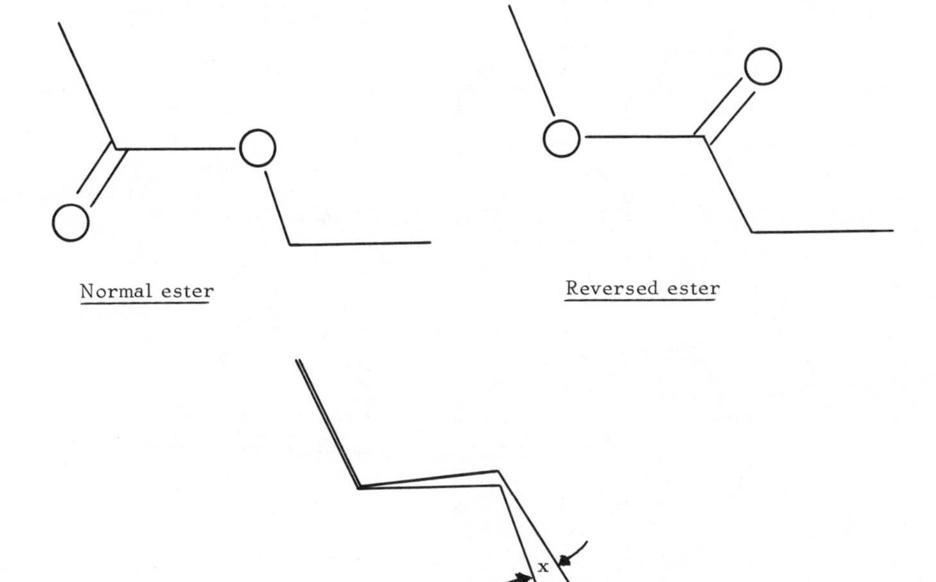

Figure 3.

The reduction in activity that accompanies reversal of the ester linkage might be ascribed to a change in the geometry of the linkage, as shown in Figure 3 (x = approximately 10°). This could alter the structural interactions of the "alcoholic" portion at the site of action. However, the results of this study demonstrate that, providing the linkage has similar spatial characteristics to the normal ester, a carbonyl group adjacent to the cyclopropane ring is not an essential requirement for pyrethroid-like activity.

Table IV.

No.	COMPOUND	ESTER LINK	LD50 γ/♀		F.o.S.
			ALONE	+16824(1:5)	
V	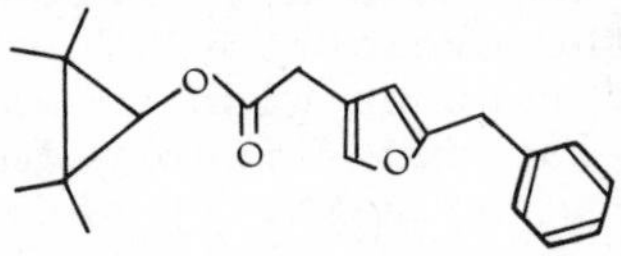	R	5.14	0.19 0.16	27 32
VI	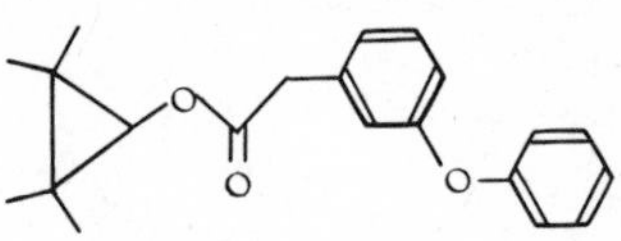	R	>6.0	2.3	>2.6
NRDC 108	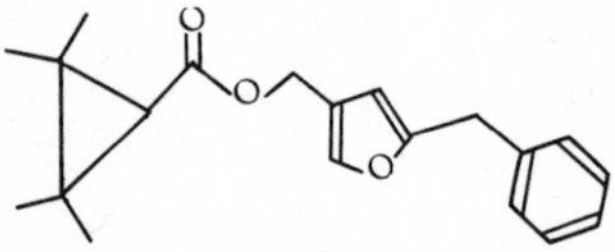	N	0.022	0.020	1
VII	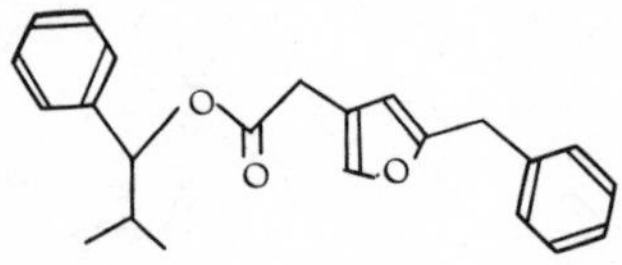	R	4.8	1.0	4.8
VIII	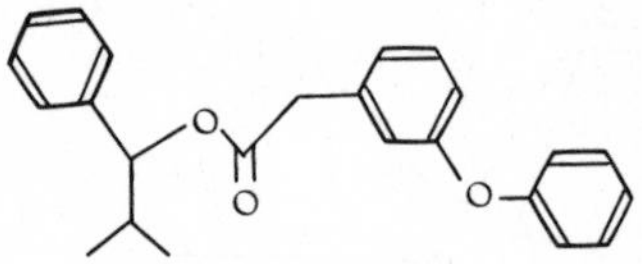	R	c6.0	2.0	3
	BIOALLETHRIN	-	0.42	0.12+	-

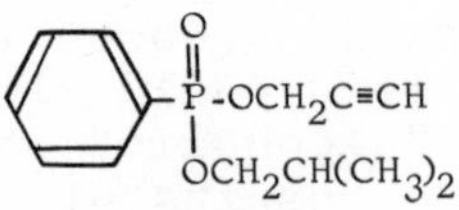

$C_6H_5-P(=O)(-OCH_2C{\equiv}CH)(OCH_2CH(CH_3)_2)$

FMC 16824

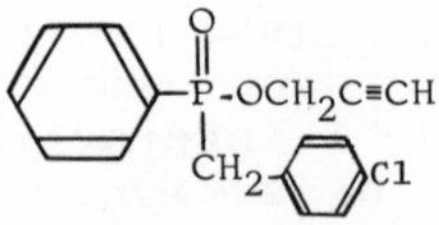

$C_6H_5-P(=O)(-OCH_2C{\equiv}CH)(CH_2-C_6H_4-Cl)$

FMC 11523

Evaluation of Insecticidal Activities.

Compounds were tested by topical application of cellosolve solutions (0.6 μl alone and 0.24 μl with piperonyl butoxide) on the dorsum of the thorax of female houseflies (*Musca domestica* - laboratory susceptible strain) by a micrometer syringe. The dose to give an LD50, based on the mortality twenty-four hours after treatment, was established. The maximum dose for compounds tested alone was 6γ and for those in combination with piperonyl butoxide (1:5) 2.4γ.

Literature Cited.

(1) Sota, K., *et al*. Botyu-Kagaku, (1973), **38**, 181.
(2) Osbond, J.M. & Wickens, J.C., German Patent Offen, 1,961,777.
(3) Elliott, M., Bull. Wld. Hlth. Org., (1971), **44**, 315-324.
(4) Baker, R.H., Martin, W.B., J. Org. Chem., (1960), **25**, 1496.
(5) Matsui, M., *et al*., Bull. Agr. Chem. Soc., Japan, (1956), **20**, 89.
(6) Katsuda, S., Japanese Patent 8498, (1961).
(7) Berteau, P.E., Casida, J.E., J. Agr. Food Chem., (1969) **17**. 931.
(8) Belgium Patent, 801,946.
(9) U.S. Patents. 3,885,031; 3,557,259.

7

Insecticidal Activities of Synthetic Pyrethroids

YOSHIYUKI INOUE, SHIGEKI OHONO, TAKAO MIZUNO,
YASUO YURA, and KEISUK MURAYAMA

Central Research Laboratories, Sankyo Co. Ltd., No. 2-58 Hiromachi 1-chome,
Shinagawa-ku, Tokyo 140, Japan

Pyrethroids (natural and synthetic ones) are becoming increasingly important as insect control agents because they possess a unique combination of desirable properties including exceptionally good insecticidal activity, low mammalian toxicity, and rapid biodegradation. These features, combined with their broad spectrum of insecticidal activities, have made them commercially successful, and also environmentally safe.

There have been desired synthetic insecticide having higher toxicity on insects, lower cost, lower mammalian toxicity and some unique properties for actual uses. In general, the uses of pyrethroids are mainly limited by high cost and by their instability for certain possible uses.

RESULTS AND DISCUSSION

It was recognized by Nakada and collaborators (1) that several compounds were isolated by pyrolysis of allethrin heated at 400°C. Two of the compounds obtained by the pyrolysis were indanone derivatives as shown in Fig. 1. These indanones showed themselves weak insecticidal activities.

The activity was measured by using first instar nymphs of the American cockroach as first screening insect. According to this method, each compounds dissolved in acetone were deposited into a 20 ml glass vial. After evaporating the solvent, 10 nymphs of the cockroach were introduced into the vial. Then the vial was covered with a plastic lid. Mortality of nymphs after 24 hours was measured.

As shown in Fig. 2 the degree of effectiveness against nymphs of the cockroach was presented by using a mark from A to E. In the case of A, 100 per cent kill of nymphs of cockroach was obtained by using 1 μgr of the compound.

These indanones obtained by pyrolysis have a

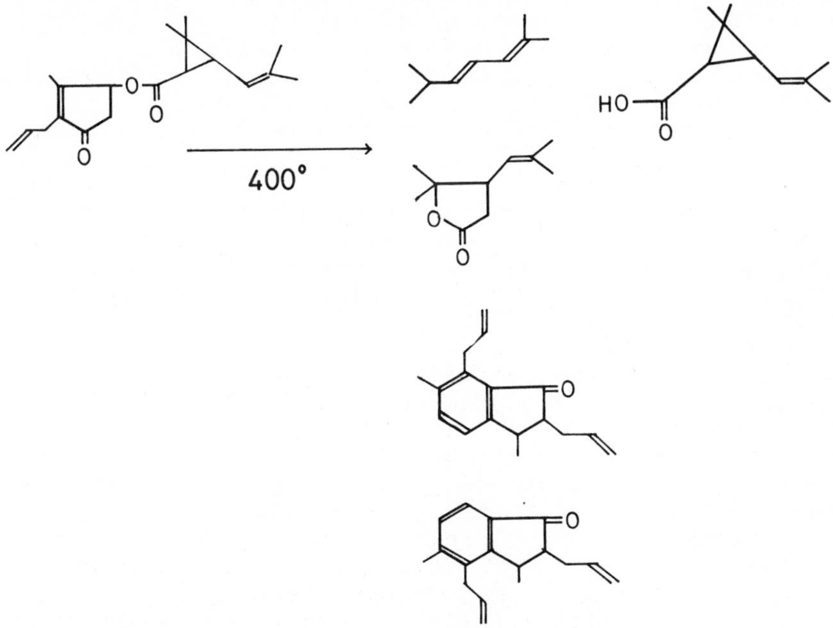

Figure 1. Pyrolysis of allethrin

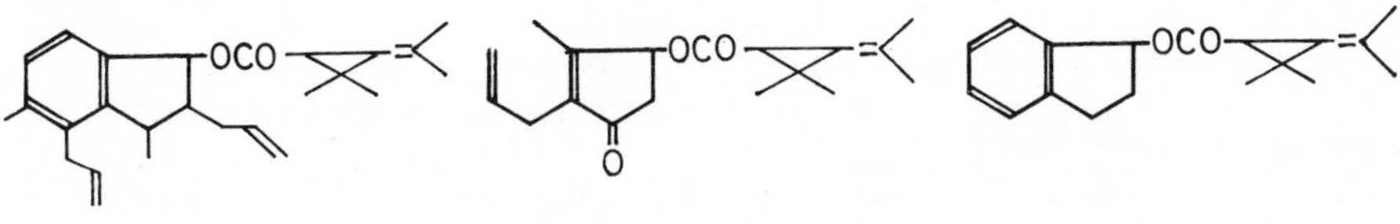

Degree of effectiveness	Insecticidal effect
A	1 μg : 100 % kill
B	10 μg : 100 % kill
C	100 μg : 100 % kill
D	1000 μg : 100 % kill
E	1000 μg : 0 %

Dose : μg / vial test insect : first instar nymphs of the American cockroach.

Figure 2. Degree of effectiveness

structural similarity to allethrolone. Thus, indanol ester of crysanthemic acid was synthesized. This compound didn't show insecticidal activity. Therefore, more simplified derivatives were synthesized.

As a result, 1-indanyl chrysanthemate indicated insecticidal activity though the activity was less than that of allethrin.

Therefore modifications of indanyl and related benzofuranyl or benzothiophenyl compounds were carried out. Chemical structure and their insecticidal activities are shown in Fig. 3. The methyl group at 4-position was effective but benzyl or chloro substituent were not effective. Further introduction of methyl group was not effective. 2,3-Dihydro-7-methyl-3-benzofuranyl chrysanthemate which call ES-56, exhibited activity.

Synthetic route of the ES-56 is shown in Fig. 4. According to our synthetic route, the first step is ether formation with o-cresol and chloroacetic acid. Conversion with thionyl chloride to the acid chloride followed by Friedel-Craft cyclization with aluminium chloride gave 7-methyl-2,3-dihydrobenzofuranone. Reduction of the ketone leads to 7-methyl-2,3-dihydrobenzofuranol. Esterification of the benzofuranol with chrysanthemic acid chloride results in final product.

The activities of ES-56 were compared with known pyrethroids.

Insecticidal activities of pyrethroids against nymphs of the cockroach and termite are shown in Table 1.

Using dryfilm method, 0.1 μgr of permethrin killed 10% of cockroach or 85% of termite. Permethrin was most effective.

The amount of 0.1 μgr of d-trans of ES-56 killed 20 per cent of cockroach.

Table 2 shows the result of insecticidal activity of pyrethroids against common housefly by topical application method. LD_{50} of ES-56 was 0.41 μgr per fly. d-trans Phenothrin and resmethrin showed higher insecticidal activity.

Table 3 shows that the insecticidal activity of pyrethroids against three species of cockroach was examined by topical application method. Only d-trans phenothrin and rethmethrin were more effective than ES-56.

The persistence of ES-56 and resmethrin are shown in Table 4 using termite, *Coptotermes formosanus*. Each amount of 0.1 mg and 1 mg of pyrethroids were added in a glass vial previously mentioned, and mortality was measured after 1 to 128 days. In the case

Table 1. Insecticidal effect of pyrethroids against nymphs of cockroach and termite by dryfilm method.

Pyrethroids	Mortality (%)									
	100 μg		10 μg		1 μg		0.1 μg		0.01 μg	
	C	T	C	T	C	T	C	T	C	T
ES-56	100	100	100	100	100	100	0	0	0	0
353 (trans of 56)	100	100	100	100	100	100	20	0	0	0
Allethrin	100	100	100	100	60	85	0	0	0	0
Phthalthrin	100	100	30	25	0	0	0	0	0	0
Resmethrin	100	100	100	100	100	100	0	30	0	0
Phenothrin(d-trans)	100	100	100	100	100	100	5	10	0	0
Permethrin(dl-cis,trans=1:1)	100	100	100	100	100	100	10	85	0	0

C : Cockroach (first instar nymphs of Periplaneta americana).
T : Termite (Coptotermes formosanus).

Table 2. Insecticidal activity of ES-56 and other pyrethroids on the housefly by topical application method.

Compounds	LD_{50} (μg/fly)
ES-56	0.41
Allethrin	1.44
Phthalthrin	1.55
Resmethrin	0.10
Phenothrin(d-trans)	0.09
Furamethrin	0.55
Proparthrin	0.59
Butethrin	0.33

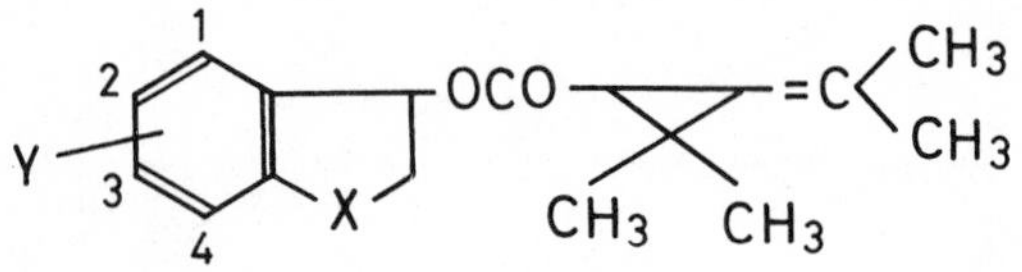

X	Substituents (Y)					
	none	1-CH_3	2-CH_3	3-CH_3	4-CH_3	4-$CH_2C_6H_5$
CH_2	C	C	C	C	B	C
O	B	—	C	B	A	B
S	B	C	D	B	B	—

Figure 3. Modifications of indanyl and related benzofuranyl or benzothiophenyl compounds

Table 3. Insecticidal activity of ES-56 and other Pyrethroids on the 3 species of cockroaches by topical application method.

Compounds	LD_{50} (μg/cockroach)		
	B.germanica	P.fuliginosa	P.americana
ES-56	2.32	20.9	18.9
Allethrin	4.75	49.0	31.1
Phthalthrin	10.70	100μg=10%	100μg=0 %
Resmethrin	1.20	15.5	7.6
Phenothrin(d-trans)	1.23	14.4	3.5
Furamethrin	6.90	21.9	63.0
Proparthrin	4.11	24.5	49.0
Butethrin	11.20	55.9	49.0

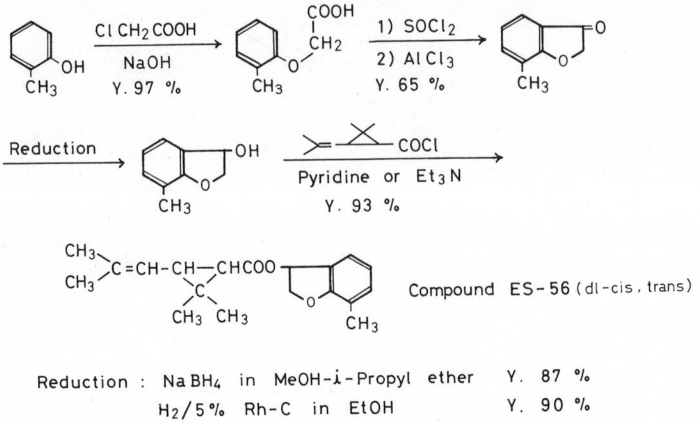

*Figure 4. Synthesis of 7-methyl-2,3-dihydro-3-benzofurylchrysanthemate (dl-*cis, trans*-mixture).*

Table 4. The persistence of effectiveness against termite.

Compounds	Dose (mg)	Mortality (%)						
		1day	7	14	28	56	64	128
ES-56	0.1	100	100	100	100	100	5	0
	1.0	100	100	100	100	100	100	0
Resmethrin	0.1	100	55	0	0	0	0	0
	1.0	100	100	100	0	0	0	0

Dry film method.

of 1 mg of ES-56, the effect to kill 100% of termites continued for 64 days. Whereas the corresponding effect of resmethrin was only 14 days.

A laboratory termite test was carried out using Coptotermes formosanus. The test block was Japanese cedar and the size of the block was 1 x 1 x 1 cm. The block was treated with methanol solution of each compounds and the volume of absorbed solution was 300 ml per m^2. The exposure to termite was carried out for 15 days without weathering. Then, weight loss of the block and survival of the termite were measured. The results are shown in Table 5. Permethrin showed the strongest effects against termite. But ES-56 was also effective.

Table 5. Results of termite test.

Prevention of attack / Concentration (%) / Pyrethroids	Weight loss (%)			Surviving (%)		
	0.2	0.02	0.002	0.2	0.02	0.002
ES-56	0	9.4	13.1	0	26	100
Permethrin (dl-cis,dl-trans,1:1)	0	0	11.1	0	22	100
Resmethrin	0	7.8	9.9	0	28	100
Phenothrin (d-trans)	0	4.4	14.4	0	30	100
Allethrin	0	18.2	23.6	0	100	100
Phthalthrin	10.4	20.6	20.6	100	100	100
Untreated		26.6			100	

Wood piece : Japanese cedar (*Cryptomeria japonica*) 10 x 10 x 10 mm.
Termite : *Coptotermes formosanus*, Exposure period : 15 days.
Solvent : methanol, Surface treatment : 300 ml/m^2

Tabe 6. Knockdown effect by vapour of pyrethroids on the housefly.

Compounds	Knockdown (%)					
	10 min.	20	30	40	50	60
ES-56	0	23.3	36.7	46.7	60.0	70.0
Allethrin	0	0	0	0	0	0
Phthalthrin	0	0	0	0	0	0
Resmethrin	0	0	0	0	0	3.3

Time (second)	60	120	180	240	300	360	420
Temperature(°C)	93	112	132	136	138	140	142

A new box type apparatus was deviced for fumigation or mist spraying test. The size of the box was 30 x 30 x 30 cm. Test insects were introduced into the box after fumigation or mist spraying without leaking of the insecticide. The diagram is shown in Fig.5.

The knockdown effect based on vapour action of pyrethroids was measured by using common housefly. One mg of each compound was heated up to 142°C of the heater during 7 minutes. Then, the test insects were introduced into the box and the percentage of knockdown was determined. The results are shown in Table 6. ES-56 showed better result than allethrin, phthalthrin and resmethrin.

The insecticidal activity of ES-56 was increased when combined with safroxane or 1-dodecylimidazole.

Using the box type apparatus previously mentioned, the effect of mist spraying to common housefly was tested using 0.4% aceton solution. The results are shown in Table 7. The knockdown effect was increased when 1-dodecylimidazole was added. The insecticidal effect of safroxane was stronger than that of 1-dodecylimidazole.

Table 7. Mist spraying test against housefly by using a new box apparatus.

Compounds	Concentration (%)	KT_{50} (min.)	Mortality (%)
Allethrin	0.4	55.2	2.5
ES-56	0.4	39.8	5.0
ES-56 + Safroxane	0.4 + 2	32.8	50.0
ES-56 + 1-Dodecyl imidazole	0.4 + 2	29.1	37.5

Solvent : acetone
Amount of spray : 0.65ml/30x30x30cm

In the case of the mixed ratio, 1 to 5, better synergistic effects were observed.

The insecticidal activity of thin wood treated with pyrethroids containing synergists was tested after exposing to ultraviolet light. In this experiment, weather -Ometer having one carbon arc lamp was used. The irradiation time was from 3 to 12 hours. After irradiation, the first instar nymphs of the American

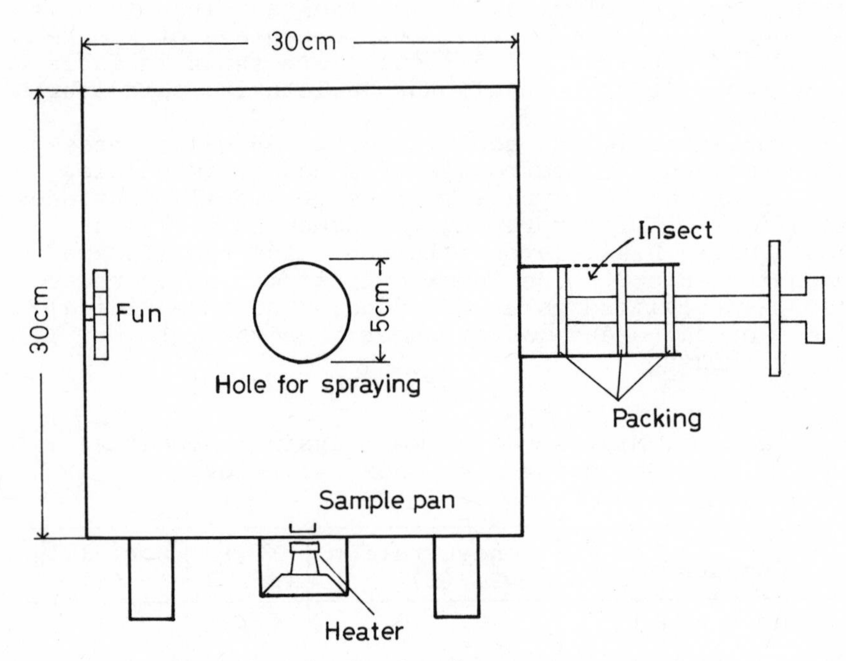

Figure 5. A diagram of new box-type testing apparatus

cockroach were contacted on the surface of the thin wood and the mortality was measured. The results are shown in Table 8.

Table 8. Insecticidal activity of wood treated with pyrethroids after exposing to ultraviolet light

Pyrethroids	Synergists	Mortality (%)				
		0hr.*	3	6	9	12
ES-56	———	100	80	30	0	0
	Safroxane	100	100	100	25	0
	1-Dodecyl imidazole	100	100	100	60	0
Resmethrin	———	100	0	0	0	0
	Safroxane	100	0	0	0	0
	1-Dodecyl imidazole	100	0	0	0	0

* Irradiation time
Concentration : 0.2% acetone solution
Mixed rate, 1 : 1
Weather-O meter was used.

ES-56 was well protected by the addition of synergists from ultraviolet light.

As a part of the toxicity studies of ES-56, LD_{50} (mg/kg) was examined. The results are shown in Table 9. The toxicity of compound 56 would be less than that of allethrin.

In conclusion, ES-56 has better insecticidal activity and relatively longer residual effect. Moreover, lower cost is to be expected. There is much of work yet to be done in fundamental tests on this new pyrethroid. We are now making extensive effort for further development of this compound, especially toxicological studies on it.

REFERENCES

1. Nakada y, Yura. Y and Murayama K : Bull. Chem. Soc. Jap. (1971), 44, 1724

Table 9. Toxicity of compound ES-56

Animals	sex	LD_{50} (mg/kg)		
		Peroral	Subcutaneous	Percutaneous
Mouse	male	692	326	3000 ~ 4000
(ddy)	female	661	334	3524
Rat	male	3590	1863	> 3 ml
(wister)	female	3412	1758	> 3 ml

Allethrin : mouse, male 513, female 416
rat, male 1084, female 1217 (Peroral)

8

Neurophysiological Study of the Structure-Activity Relation of Pyrethroids

T. NARAHASHI, K. NISHIMURA, J. L. PARMENTIER, and K. TAKENO

Department of Physiology and Pharmacology, Duke University Medical Center, Durham, N. C. 27710

M. ELLIOTT

Rothamsted Experimental Station, Harpenden, Hertfordshire, AL5 2JQ, England

Much information is now available on the relationship between the structure and insecticidal activity of various groups of insecticides, including DDT and its derivatives, organophosphates, carbamates and pyrethroids (1). A common procedure in such studies is to compare activities of compounds with systematically altered structures. To interpret such results rationally, the complex nature of the toxic action of insecticides, illustrated diagrammatically in Figure 1 (2, 3) must be recognized.

The first step in the action of an insecticide is penetration into the insect body, via the cuticle, mouth or respiratory system. The insecticide that enters will migrate to various tissues by the open circulation system. Some of the insecticide may be detoxified before reaching the target site; mixed function oxidases, for example, are known to be involved in the metabolic degradation of a variety of organophosphates, carbamates and DDT analogues. With some compounds metabolic products are more toxic than the original insecticides; their formation is termed "activation". For example, oxidation of parathion to paraxon increases potency to inhibit cholinesterases by a factor of 10^5.

Eventually, either the original or the activated insecticide reaches and influences the target site, usually the nervous system. The symptoms of poisoning are a variety of secondary and tertiary disturbances in the insect, and death finally results from these integrated toxic actions. Unlike mammals, insects do not die by a single dysfunction of a key organ. Mammals, for example, suffer respiratory failure or cardiac arrest following intoxication by the insecticide, whereas the death of insects involves a complex series of reactions in various organs such as metabolic exhaustion and paralysis of the entire nervous system.

This outline of insecticidal action indicates that the relative insecticidal activity of different compounds is the outcome of a complex series of interacting processes. Comparison of overall potencies, therefore, cannot elucidate fully structure-activity relationships, but must be supplemented by studies on primary activities at the target for thorough clarification. This approach has indeed been used widely in studying organophosphates and carbamates (4, 5) which inhibit cholinesterases. With these groups, such experiments are technically straightforward, since cholinesterases can be handled *in vitro*. However, for insecticides which do not inhibit cholinesterases the situation is more complicated; a target site preparation permitting many experiments in a short time with minimum expense is required. As discussed, most insecticides affect the nervous system, so the best model would be a nerve preparation. Few such studies with nerve preparations have been made. However, attempting to define structure-activity relationships for rotenoids Fukami *et al.* (6) compared rotenone derivatives for their potency against insects, and for ability to inhibit glutamic dehydrogenase, and to block nervous conduction. Relative effectiveness was similar for the three actions except for a few derivatives which showed weak insecticidal action despite strong inhibition of enzymic action and blocking of nervous conduction. For five synthetic pyrethroids Berteau *et al.* (7) found a good correlation between insecticidal potency, mammalian toxicity and blocking of nervous conduction. Recently Burt and Goodchild (8, 9, 10) using a sucrose gap technique tested the effects of a large number of synthetic pyrethroids on giant fibres and cervical nerve-giant fibre synapses of the cockroach, *Periplaneta americana* L. They compared the neurotoxicities with the action of the compounds on living insects, concluding that although a rational pattern of relationships was apparent for overall toxicity, no comparable connection could be discerned for neurotoxicity, except that neurotoxicity tended to increase with polarity. Neither site of action was likely to contain a critical site of action for pyrethroids.

Interesting results were also obtained with DDT analogues (3, 11, 12). Although insecticidal potency correlated well with ability to increase the negative (depolarizing) after-potential and to induce repetitive after-discharges for most of the derivatives tested, striking anomalies were found with other derivatives. For example, substituting amino or hydroxy groups for the p,p'-chlorines of p,p'-DDT made the compound insecticidally inactive (13), yet the analogues were still active on the nerve (12), but in a manner, blocking rather than excitatory. Thus their action is entirely different from that of the parent compound, p,p'-DDT. The structure-activity relationship can therefore only be fully defined by experiments using the target site *in vitro*.

Methods

A simple method has been developed whereby potency to affect the nervous system of a large number of compounds can be

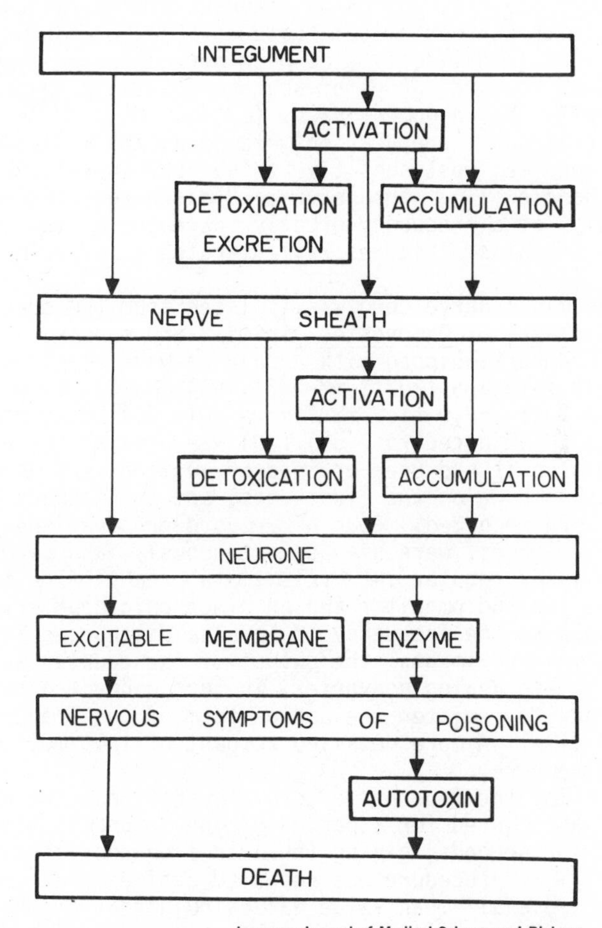

Figure 1. Process of toxic action of an insecticide (2)

compared (14). Natural pyrethroids and allethrin are known to stimulate and then paralyze various nerve preparations (15, 16, 17, 18, 19, 20, 21).

The compounds used in this work were prepared as described previously (22, 23, 24, 25) or by standard procedures, and their structures, with stereochemical features (26), are shown in Table I. Insecticidal activities were determined by topical application of measured drops of acetone solutions to the adult, female housefly, Musca domestica L. (27).

Of several nerve preparations examined, the isolated crayfish nerve cord was most sensitive to various insecticides and easiest to handle (14). The nerve cord discharges impulses spontaneously, at a frequency greatly increased by low concentrations of pyrethroids. Figures 2 illustrates an experiment with allethrin.

The abdominal nerve cord was isolated from the crayfish, Procambarus clarki or Orconestes virilis, and mounted in a Plexiglass chamber equipped with a pair of wire electrodes. van Harreveld solution (207.3 mM NaCl, 5.4 mM KCl, 13.0 mM $CaCl_2 \cdot 2H_2O$, 2.6 mM $MgCl_2 \cdot 6H_2O$, 1.9 mM Trizma HCl, 0.4 mM Trizma Base with a final pH adjusted to 7.55) (28) was used as the bathing medium, and when it had been drained by suction, the nerve cord preparation was hung on the electrodes, and spontaneous impulse discharges were recorded. Four nerve cord preparations, mounted in separate chambers, were used simultaneously, and each switched electronically at regular intervals via a preamplifier to an oscilloscope, an audiomonitor and an electronic counter. The counter displayed the frequency of impulse discharges in digital form. In some experiments, the output of the counter was fed into a digital-to-analog converter, in turn connected to a strip chart recorder to register the analog form of the frequency as a function of time. A more detailed account of this method will be published elsewhere.

For each of the four nerve cord preparations, the number of discharges was counted for a period of one second 15 times at an interval of 1-2 seconds, giving the mean frequency of discharges per second. This procedure was repeated 3 times every 10 minutes, and the overall mean value was calculated from the 45 measurements. Then the lowest concentration of a test compound (usually 1×10^{-8} M) prepared from a stock solution in ethanol, was applied to the nerve. Frequency counts were made 10, 20 and 30 minutes after applying the test compound. After the last count, the concentration of the test compound was increased 10-fold, and three sets of counts were made every 10 minutes. These procedures were repeated until the concentration of the test compound reached 1×10^{-5} M, the highest value tested. The ethanol concentration in the test solution was 0.1% (v/v) at the highest test compound concentration of 1×10^{-5} M, and had no effect on the spontaneous discharges of the nerve cord.

Dose-response curves were constructed by plotting overall

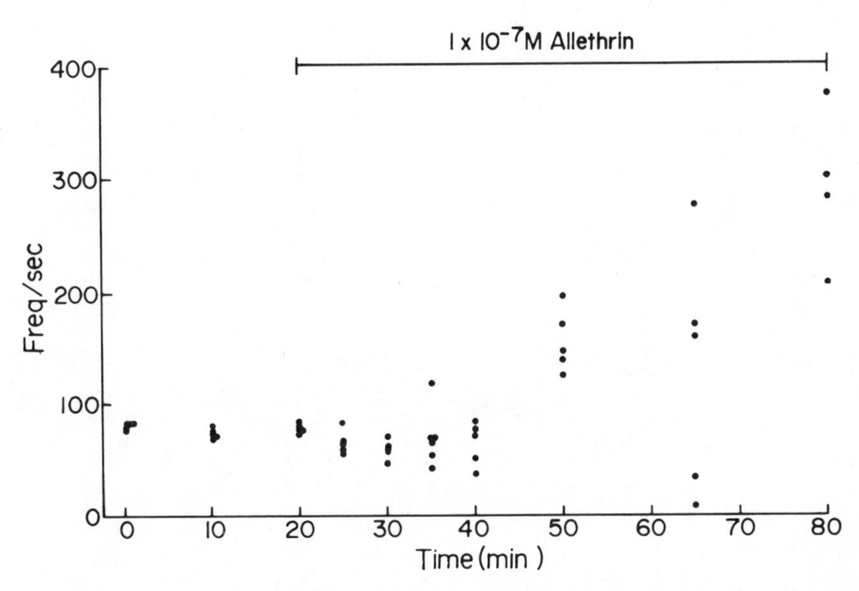

Figure 2. Frequency of impulse discharges from the abdominal nerve cord of the crayfish before and during application of allethrin at a concentration of 1×10^{-7} M

Table I. The Synthetic Pyrethroids Used

Compound No.	Alcoholic Component	Acidic Compound
1	3-phenoxybenzyl	[1R, _trans_]-chrysanthemate
2	6-cyano-3-phenoxybenzyl	[1R, _trans_]-chrysanthemate
3	6-chloro-3-phenoxybenzyl	[1R, _trans_]-chrysanthemate
4	4-benzoyloxybenzyl	[1R, _trans_]-chrysanthemate
5	3-phenoxymethylbenzyl	[1R, _trans_]-chrysanthemate
6	3-phenoxy-4-methylbenzyl	[1R, _trans_]-chrysanthemate
7	7-phenoxy-tetrahydronaphth-1-yl	[1R, _trans_]-chrysanthemate
8	α-cyano-3-phenoxybenzylamide of	[1R, _trans_]-chrysanthemic acid
9	3-phenoxybenzyl	[1R, _cis_]-chrysanthemate acid
10	5-benzyl-3-furylmethyl	[1R, _trans_]-2,2-dimethyl-3-(2,2-difluorovinyl)cyclopropane-carboxylate
11	5-benzyl-3-furylmethyl	[1R, _cis_]-2,2-dimethyl-3-(2,2-difluorovinyl)cyclopropane-carboxylate
12	3,4,5,6-tetrahydrophthalimidomethyl	[1R, _cis_]-2,2-dimethyl-3-(2,2-difluorovinyl)cyclopropane-carboxylate
13	(±)-α-cyano-3-phenoxybenzyl	[1R, _trans_]-2,2-dimethyl-3-(2,2-dichlorovinyl)cyclopropane-carboxylate
14	3-phenoxybenzyl	[1S, _trans_]-2,2-dimethyl-3-(2,2-dichlorovinyl)cyclopropane-carboxylate
15	5-benzyl-3-furylmethyl	[1RS, _cis_]-2,2-dimethyl-3-(2,2-dichlorovinyl)cyclopropane-carboxylate
16	α-cyanopiperonyl	[1RS, _cis_, _trans_]-2,2-dimethyl-3-(2,2-dichlorovinyl)cyclopropane-carboxylate
17	5-benzyl-3-furylmethyl	[1RS, _cis_, _trans_]-3-methyl-3-isobutenylcyclopropanecarboxylate

Table I (continued)

Compound No.	Alcoholic Component	Acidic Compound
18	5-benyzl-3-furylmethyl	(+)-α-isopropyl-2-fluorophenylacetate
19	5-benzyl-3-furylmethyl	(±)-α-isopropyl-4-isopropylphenylacetate
20	5-benzyl-3-furylmethyl	2-naphthoate
21	5-benzyl-3-furylmethyl	4-tertiarybutylbenzoate
22	α-cyano-3-methoxybenzyl	[1R, <u>trans</u>]-chrysanthemate
23	3-phenoxybenzyl	α,α-diethyl-4-chlorophenylacetate
24	3-phenoxybenzyl	(+)-α-isopropyl-3,5-dimethylphenylacetate
25	3-phenoxybenzyl	(±)-α-isopropyl-3,4-dimethylphenylacetate
26	3-phenoxybenzyl	(±)-α-isopropyl-4-ethylphenylacetate
27	3-phenoxybenzyl	(±)-α-isopropyl-2-fluorophenylacetate
28	3-phenoxybenzyl	(±)-α-isopropyl-3-methylphenylacetate
29	3-phenoxybenzyl	(±)-α-methyl-α-isopropyl-4-chlorophenylacetate
30	3-phenoxybenzyl	(±)-α-cyclohexyl-4-chlorophenylacetate
31	3-phenoxybenzyl	(±)-indan-1-carboxylate
32	3-phenoxybenzyl	3-methyl-3-(4-methylphenyl)-butyrate
33	3-phenoxybenzyl	(+)-6-chloro-2-methyl-1,2,3,4-tetrahydronaphthoate
34	3-phenoxybenzyl	(±)-α-cyclopentyl-4-chlorophenylacetate
35	3-phenoxybenzyl	(±)-2,2-dichloro-3,3-dimethylcyclopropanecarboxylate
36	3-phenoxybenzyl	(±)-α-dimethylamino-4-chlorophenylacetate
37	3-phenoxybenzyl	phenyl-isopropylcarbamate

mean values for 45 counts in the control and in each concentration of a test compound against the logarithm of the concentration. The frequency of impulse discharges passed through a maximum with increasing concentration. By connecting each measurement by a straight line, the concentration at which the frequency increased to 200% of the control was estimated, and designated NS_{200} (nerve stimulation to 200%). Some compounds did not stimulate the nerve to increase the impulse frequency to 200% of the control even at 1×10^{-5} M.

Burt and Goodchild (19) examined the sensitivity of the terminal ganglion of the abdominal nerve cord of *Periplaneta americana* to pyrethrin I by a method similar in principle and obtained comparable results.

Nerve Action Vs. Insecticidal Action

The synthetic pyrethroids differed greatly in their ability to stimulate spontaneous impulse discharges of the crayfish abdominal nerve cord. The value of NS_{200} is plotted against the lethal dose 50 (LD_{50}) relative to that of bioresmethrin (0.005 ng/insect) (Figure 3). If nerve potency alone determined insecticidal activity, all measurements would fall on a line with a definite slope. However, many of the compounds deviated greatly from such a simple relationship. For example, compounds 1, 9, 13 and 15 were approximately equally toxic to insects, yet their nerve stimulating potencies were greatly different, the NS_{200} ratio of 15 to 9 being more than 10^3. Correspondingly, variations of nerve potency in compounds with similar insecticidal activity were observed with the compound 14 which had a high NS_{200} of 1.5×10^{-8} M and 37 and four other compounds which did not stimulate the nerve at 1×10^{-5} M. Thus the difference is nerve potency between these two groups is over 600.

Some compounds had approximately the same nerve potency, yet differed considerably in their insecticidal potency. For example, the compounds 10 and 23 had comparable NS_{200} values, but the former was 200 times effective as an insecticide. Likewise, the compound 9 was almost equipotent to 14 in respect to the nerve action, yet the former was about 50 times more effective as an insecticide. The compound 13 was one-tenth as potent on the nerve than 14, yet 50-fold more active insecticidally.

Comparison of Isomers and Analogs

Very interesting differences in activity were disclosed by comparing isomers and analogs for their effects on the nerve. For example the 5-benzyl-3-furylmethyl (+)-*cis*-fluorovinyl ester, 11, was highly potent on the nerve with a NS_{200} value of 1.75×10^{-9} M, whereas the corresponding (+)-*trans* isomer, 10, was 53-fold less effective with a NS_{200} value of 9.2×10^{-8} M. However, the insecticidal potencies were less drastically

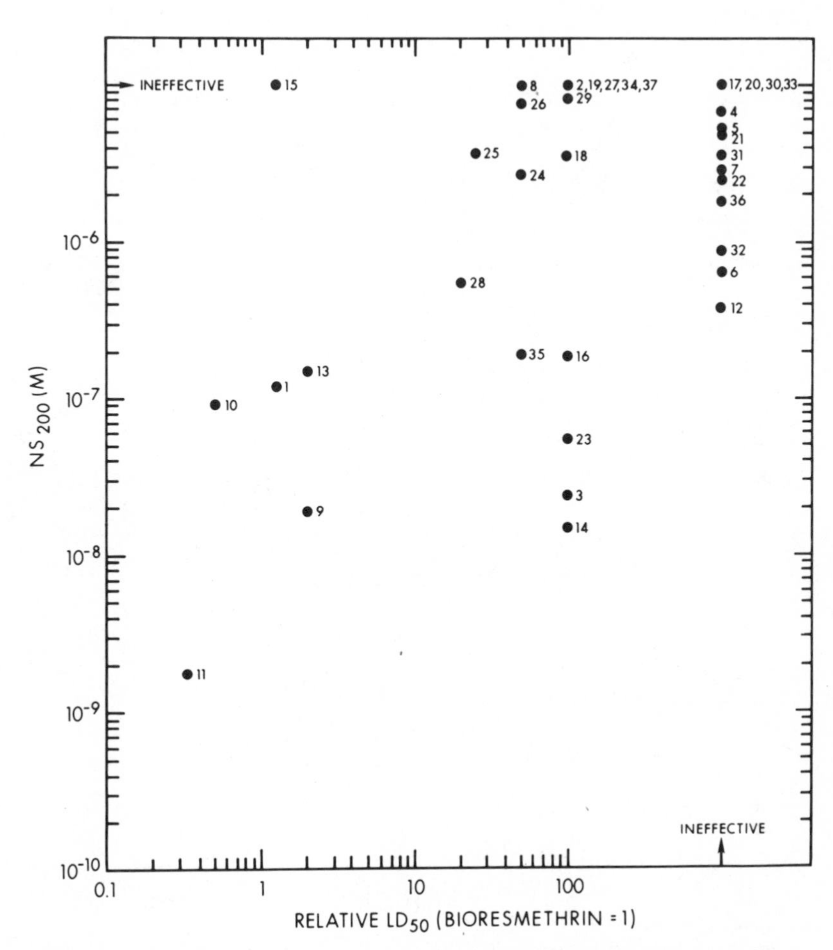

Figure 3. Relationship between the concentration to stimulate the impulse frequency of the crayfish abdominal nerve cord to 200% of the control (NS_{200}) and the lethal dose 50 (LD_{50}) against the housefly for synthetic pyrethroids

Table II. Comparison of 5-Benzyl-3-furylmethyl Pyrethroids for Their Structures, NS_{200} and LD_{50} Values

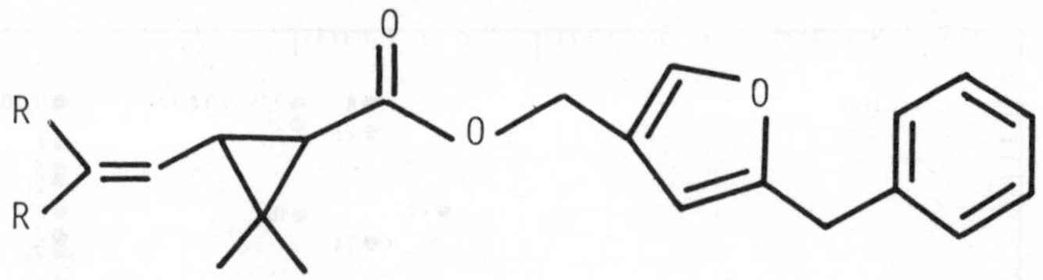

Compound No.	Stereochem-istry	R	NS_{200}* ($X\ 10^{-8}$ M)	Relative* LD_{50}
11	[1R, _cis_]	F	0.175	0.33
10	[1R, _trans_]	F	9.2	0.5
15	[1RS, _cis_]	Cl	-----†	1.25
Bioresmethrin	[1R, _trans_]	CH_3	5	1

*Lowest value corresponds to highest potency.

†No stimulating effect at 1×10^{-5} M.

Table III. Comparison of 3-Phenoxybenzyl Pyrethroids for Their Structures, NS_{200} and LD_{50} Values

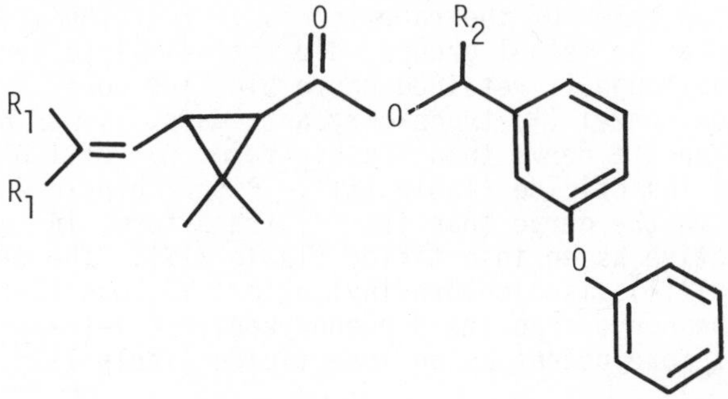

Compound No.	Stereochemistry	R_1	R_2	NS_{200}* ($\times 10^{-8}$ M)	Relative LD_{50}*
1	[1R, trans]	CH_3	--	12	1.25
9	[1R, cis]	CH_3	--	1.9	2
Biopermethrin (NRDC 147)	[1R, trans]	Cl	--	--†	0.6
14	[1S, trans]	Cl	--	1.5	100
13 (NRDC 168)	[1R, cis]	Cl	CN	15	2

*Lowest value corresponds to highest potency.

†No stimulating effect at 1×10^{-5} M.

different, the (+)-cis form being more potent than the (+)-trans form by a factor of only 1.5 (Table II). Again, 5-benzyl-3-furylmethyl (+)-trans-chrysanthemate (bioresmethrin) showed high nerve and insecticidal potencies, but the equally active insecticide, 15, an ester of the racemic cis form of the acid with chlorine replacing methyl groups, did not stimulate nerves (Table II) although it retained nerve blocking power.

3-Phenoxybenzyl (+)-trans-chrysanthemate, 1, was 6.3-fold less potent on the nerve than its cis form, 9, yet 1.6-fold more potent as an insecticide (Table III). Biopermethrin was much less potent on the nerve than its (-)-trans form, 14, yet 167-fold more active as an insecticide (Table III). The α-cyano-3-phenoxybenzyl (+)-cis-dichlorovinyl ester, 13, was 10-fold less potent on the nerve than the 3-phenoxybenzyl (-)-trans ester, 14, but 50 times more potent as an insecticide (Table III).

Conclusions

Most of the insecticidally active pyrethroids stimulated the isolated crayfish abdominal nerve cord, increasing the frequency of impulse discharges. Such stimulating action paralleled ability to kill insects with some of the compound, but others, although potent on the nerve, were weak insecticides, and vice versa. Some of the discrepancies may be due to differential cuticle penetration and detoxication, but these factors do not adequately account for all the differences (for example, those between the 5-benzyl-3-furylmethyl (+)-trans- and (+)-cis-difluorovinyl isomers, 10 and 11) and anomalies (such as the lack of nerve stimulating activity of the potent insecticide 5-benzyl-3-furylmethyl (+)-cis-dichlorovinyl ester, 15, and the inversion in activity of the (+)- and (-)-trans isomers of permethrin). The results indicate that properties and activities, other than those considered and evaluated in the present work, may be more directly related to insecticidal action of some pyrethroids.

Acknowledgements. This study was supported by a grant from the National Institutes of Health (NS06855). Technical assistance from Pamela Van Buskirk and secretarial assistance from Virginia Arnold and Arlene McClenny are greatly appreciated. Michael Elliott thanks his colleagues in the Department of Insecticides and Fungicides, Rothamsted Experimental Station, for preparing and evaluating most of the compounds used and for many valuable discussions and comments.

Literature Cited

1. Metcalf, R. L. and McKelvey, J. J., Jr. "The Future for Insecticides. Needs and Prospects, 524 pp., John Wiley & Sons, New York, 1976.

2. Narahashi, T., Jap. J. Med. Sci. Biol. (1964), 17, 46.
3. Narahashi, T., Effects of insecticides on excitable tissues. In Beament, J. W. L., Treherne, J. E. and Wigglesworth, V. B., "Advances in Insect Physiology", Vol. 8, p. 1-93, Academic Press, London and New York, 1971.
4. Fukuto, T. R., Bull. World Health Org. (1971), 44, 31.
5. Metcalf, R. L., Bull. World Health Org. (1971), 44, 43.
6. Fukami, J., Nakatsugawa, T. and Narahashi, T., Jap. J. Appl. Entom. Zool. (1959), 3, 259.
7. Berteau, P. E., Casida, J. E. and Narahashi, T., Science (1968), 161, 1151.
8. Burt, P. E. and Goodchild, R. E., Rothamsted Experimental Station, Ann. Rep. (1975) (Part 1), 155.
9. Burt, P. E. and Goodchild, R. E., Rothamsted Experimental Station, Ann. Rep. (1976), in press.
10. Burt, P. E. and Goodchild, R. E., Pesticide Sci. (1977), in press.
11. Yamada, M. and Narahashi, T., Bull. Entom. Soc. Amer. (1968), 14, 208.
12. Wu, C. H., van den Bercken, J. and Narahashi, T., Pesticide Biochem. Physiol. (1975), 5, 142.
13. Metcalf, R. L. and Fukuto, T. R., Bull. World Health Org. (1968), 38, 633.
14. Narahashi, T., Environmental Health Effects Research Series (1976), (EPA-600/1-76-005, EPA, Research Triangle Park, N. C.).
15. Lowenstein, O., Nature (1942), 150, 760.
16. Narahashi, T., J. Cell. Comp. Physiol. (1962), 59, 61.
17. Welsh, J. H. and Gordon, H. T., J. Cell. Comp. Physiol. (1947), 30, 147.
18. Yamasaki, T. and Ishii, T., Oyo-Kontyu (J. Nippon Soc. Appl. Entom.), 7, 157.
19. Burt, P. E. and Goodchild, R. E., Entomol. Exp. Appl. (1971), 14, 179.
20. Camougis, G. and Davis, W. M., Pyrethrum Post (1971), 11, 7.
21. Camougis, G., Mode of action of pyrethrum on arthropod nerves. In Casida, J. E., "Pyrethrum", 211-222, Academic Press, New York and London, 1973.
22. Elliott, M., Farnham, A. W., Janes, N. F., Needham, P. H. and Pulman, D. A., Nature (1973), 244, 456.
23. Elliott, M., Farnham, A. W., Janes, N. F., Needham, P. H., Pulman, D. A. and Stevenson, J. H., Nature (1973), 246, 169.
24. Burt, P. E., Elliott, M., Farnham, A. W., Janes, N. F., Needham, P. H. and Pulman, D. A., Pesticide Sci. (1974), 5, 791.
25. Elliott, M., Farnham, A. W., Janes, N. F., Needham, P. E. and Pulman, D. A., Pesticide Sic. (1975), 6, 537.
26. Elliott, M., Janes, N. F. and Pulman, D. A., J. Chem. Soc. Perkin I (1974), 2470.
27. Farnham, A. W., Pesticide Sci. (1973), 4, 513.
28. van Harreveld, A., Proc. Soc. Exp. Biol. Med. (1936), 34, 428.

9

Central vs. Peripheral Action of Pyrethroids on the Housefly Nervous System

T. A. MILLER and M. E. ADAMS

Department of Entomology, University of California, Riverside, Calif. 92502

Despite years of research, the site and mode of action of pyrethroids have defied description, even the simplest classification as peripheral or centrally acting has not been possible to date. On one hand, pyrethroids act in a manner resembling DDT, which is known to be a peripheral neurotoxin in insects (1). The activities of both DDT (1) and pyrethrum (2) exhibit a negative dependence on temperature, being more toxic at lower temperature. Both DDT and pyrethroids produce negative after potentials, and repetitive discharge to single stimuli in axons (3). And both DDT and pyrethrum are extremely sensitive in causing trains of sensory nervous impulses when perfused on leg preparations of insects (4, 5, 6). The actions of DDT and allethrin are also similar on the lateral-line organ of the clawed toad, *Xenopus laevis* (7).

Despite the impressive actions of pyrethroids on sensory nerve structures and the similarity between the actions of DDT and pyrethroids on isolated preparations on the nervous system, there is evidence of actions by pyrethroids on the central nervous system. Burt and Goodchild (8) found that speed of knockdown was proportional to the distance between the site of topical application and the central nervous system. They considered this to suggest strongly that knockdown, even the rapid knockdown reported by Page and Blackith (9), is due to an action on the central nervous system.

Burt and Goodchild (10) found that the isolated and perfused central nervous system of *Periplaneta americana* was sensitive to extremely low concentrations of pyrethrin I (below $5 \times 10^{-8}\underline{M}$). In contrast, DDT was without effect on the thoracic ganglia of *Periplaneta* even when applied in emulsions of $4.5 \times 10^{-3}\underline{M}$ concentration (6). This latter observation is the best demonstration of a difference between the actions of DDT and pyrethroids.

The ultimate actions of pyrethroids, then, could involve central and peripheral nervous structures--which of these might be involved during poisoning has been difficult to show until *in vivo* recording methods were developed to record the activity of flight motor units of the intact house fly during poisoning (11, 12, 13, 14, 15). This preparation allows the monitoring of a central neural coordination between flight motor neurons from the whole intact house fly during poisoning. Using this method, it was found that the pattern of activity recorded during DDT poisoning was unique and distinct from that of centrally acting poisons (11, 12). Furthermore, since the pyrethroid *trans*-Barthrin showed patterns of activity similar to DDT and unlike central nervous poisons, it was concluded that *trans*-Barthrin was peripherally acting (12).

These results were challenged informally by Paul Burt (16), so that a more extensive examination was undertaken to characterize the actions of pyrethrins.

Methods

Three pyrethroids were chosen for extensive examination. Tetramethrin (synonyms=neopynamin, phthalthrin, 2,3,4,5-tetra= hydrophthalimidomethyl chrysanthemate) was said to have very fast knockdown but poor toxicity on insects (17). We obtained a sample from Richard Hart, Wellcome Research Laboratories, Berkhamsted. This sample was reportedly a ±25/75 *cis/trans* chrysanthemate mixture.

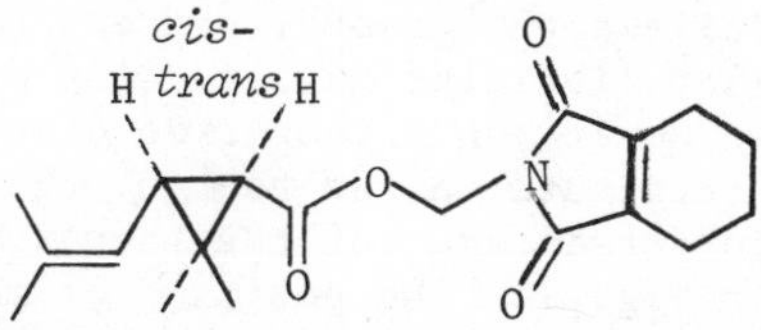

cis- H *trans* H O O O N O tetramethrin

Cis-methrin was chosen for intermediate kill and knockdown properties (NRDC 119). *Cis*-methrin is 5-Benzyl-3-furylmethyl (+)*cis* chrysanthemate:

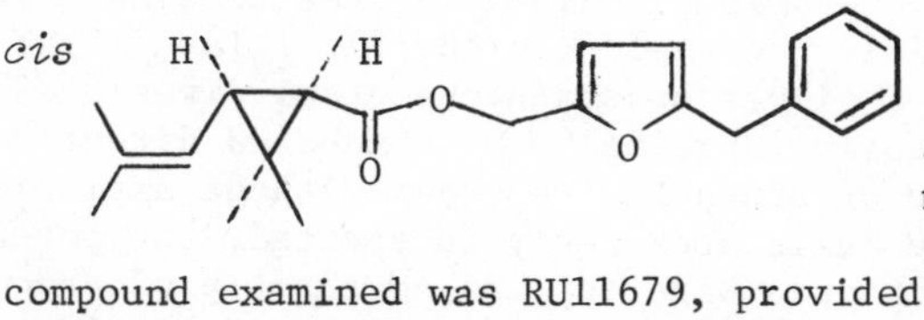

cis H H O O O *cis*-methrin

The third compound examined was RU11679, provided by the Procida Chemical Company through Wellcome Research (synonyms=k-Othrin, Bioethanomethrin) 5-Benzyl-3-furylmethyl (+)*trans* ethanochrysanthemate.

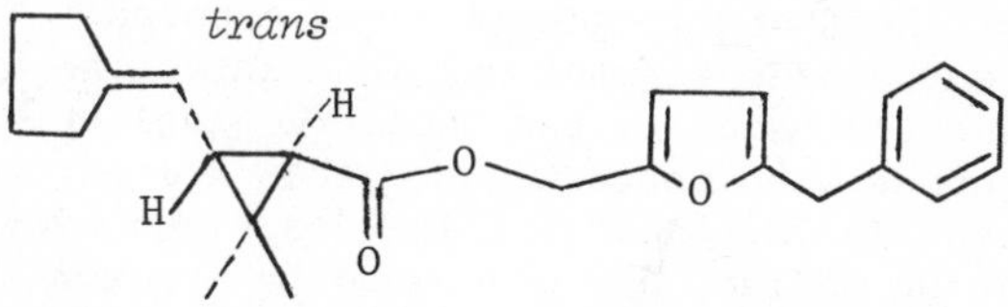

trans H H O O O

For purposes of comparison, DDT was used as well as the carbamate insecticide carbofuran (2,2-dimethyl-2,3-dihydrobenzo=furanyl-7 *N*-methyl carbamate).

Physiological preparations included the house fly leg, the cockroach leg, the flight motor of the house fly in both intact and dissected preparations. Cockroach leg preparations were essentially those developed by Roeder and Weiant (5, 6).

Nervous impulses ascending the crural nerve of the cockroach, *Periplaneta americana*, were recorded according to ordinary procedures using a grease electrode (18). Exposed tissues were bathed in a carbonate saline at pH 6.9 (19). Compounds were dissolved in acetone, then diluted at least 20 times into saline. The solution was injected into the tibia through the opening left after removal of the tarsus. Enough solution was injected to displace the hemolymph in the leg.

Nervous impulses were recorded from the isolated metathoracic leg of the house fly. After removal at the coxa-femur joint, the leg was stapled to wax as above. An electrolytically etched tungsten wire 1 mil (25 μm) in diameter was punched through the cuticle of the tibia near the femur. An indifferent electrode was placed in the femur. Compounds were applied topically in 0.1λ droplets of acetone which was just enough to wet the tarsus.

The flight motor preparation has been described in detail elsewhere (12, 13). The arrangement of dorsolongitudinal flight muscles in *Musca domestica* follows the general pattern for muscoid flies (20). Six pairs of giant fibrillar muscle cells occupy a large portion of the thoracic cavity on either side of the mid line (Fig. 1). The muscles originate on the rear of the thoracic box and insert on the anterio-dorsal area of the thorax. The area of insertion is shown in Fig. 1. The position of major bristles on the dorsal thorax allows muscle insertions to be located quite accurately and electrode wires are placed just through the cuticle over the muscles of interest and waxed in place. Only dorsolongitudinal muscles were used in this study.

Potentials recorded from the fibrillar flight muscles with non-insulated silver or stainless steel wires of 25 μm diameter are sufficiently large as to be connected directly to monitoring oscilloscope or Brush 220 recorder without amplification. Since these muscle cells are singly innervated, the potentials are simple. Compounds were applied to the tip of abdomen in acetone.

For measurements of the inherent toxicity of pyrethroids on the thoracic ganglion of house flies, females were implanted with wire electrodes in the appropriate muscles, then mounted upside down. The front two pairs of legs were removed and the ventral surface of the thoracic ganglion was exposed. The furcasternite of the mesothoracic segment was left intact as its removal caused excessive disturbance to the thoracic ganglion.

Compounds were perfused onto the exposed thoracic ganglion usually once in *Calliphora* saline described by Berridge (21) as IV-dissecting medium, and also used by Thomson (22) for *Phormia*.

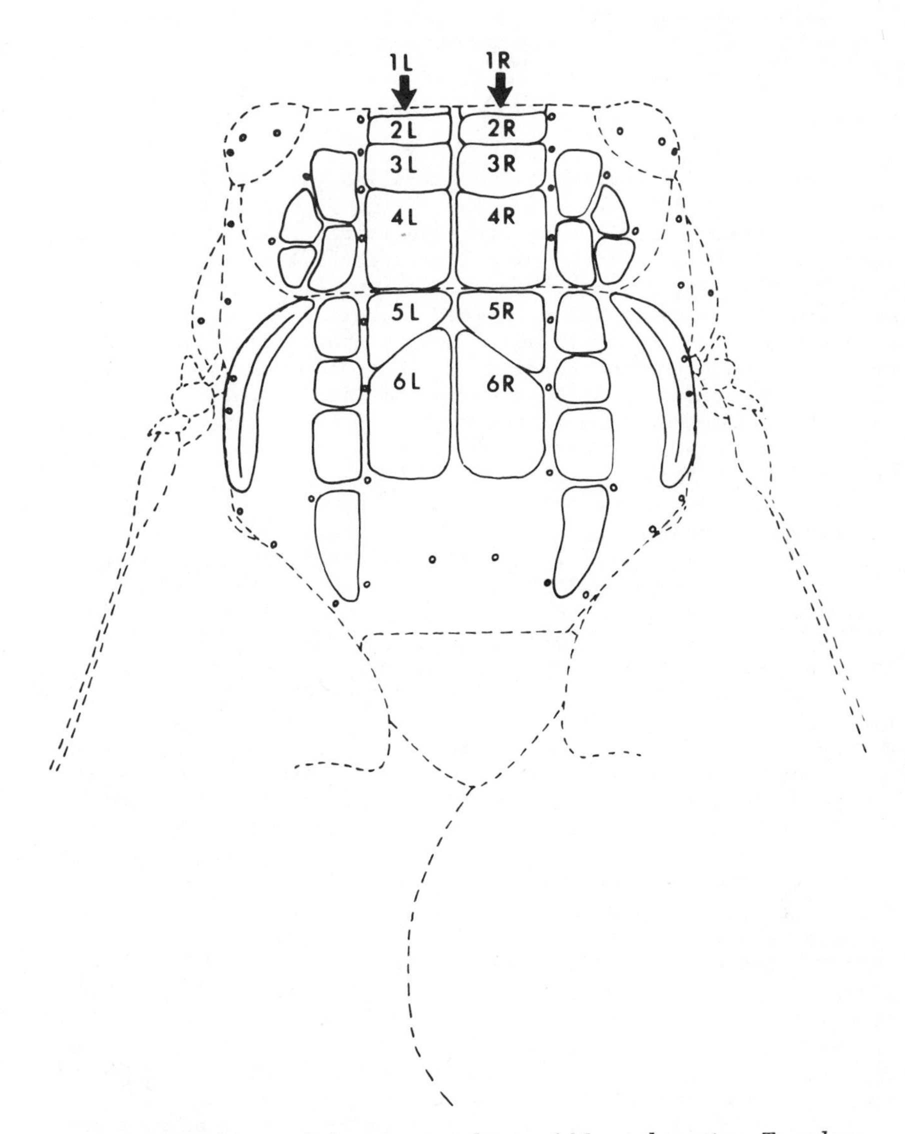

Figure 1. Dorsal view of thoracic musculature of Musca domestica. *Two dorso-longitudinal flight muscles (DLM) run longitudinally in the dorsal thorax, each comprised of six fibers. The six fibers of each DLM are stacked unilaterally and are designated 1–6 from ventral to dorsal. Recordings were made by inserting fine wires just below the cuticle into the appropriate motor unit. The desired unit was located in relation to the major bristles depicted in the map with circles.*

The saline pH was adjusted to 6.9 and rechecked after addition of compounds. The procedure was described briefly in a previous study (14).

Results and Discussion

There are 2 dorsolongitudinal muscles (DLM) in *Musca domestica*, each containing 6 fibers (Fig. 1). Each of the 12 fibrillar muscle fibers are sometimes called giant fibers since they are single multinucleate cells. Fibers 1 through 4 each have a single separate motor neuron innervating them. The two dorsal-most fibers in the right DLM (labelled 5R and 6R in Fig. 1) are innervated by a single motor neuron whose axon bifurcates before reaching the muscle fibers. The same pattern of innervation is true of the muscle cells in the left DLM.

The 5 motor neurons in the right DLM innervating the 6 muscle cells comprise 5 motor units. A motor unit is one neuron and each muscle cell it innervates.

In *Calliphora*, each of the 5 motor neurons in a DLM is functionally connected such that an antidromic nerve impulse in one unit can reset the firing rhythm of a neighboring unit (23). This functional connection is thought to be a strong lateral inhibitory connection between any one neuron and the other 4 of the DLM (23, 24). It is assumed that *Musca* has similar central nervous connections between flight motor neurons.

Antidromic impulses ascending axons of any motor unit of the right DLM have no resetting influence on the units of the left DLM in *Calliphora* (23). A similar condition likely occurs in *Musca* and implies that the units of the right DLM are weakly connected to units of the left DLM. The nature of these connections is thought to involve lateral inhibitory innervation (23); however, the details of circuitry are not known at present.

For purposes of analyzing the action of neuro-active insecticides, it is sufficient to emphasize only a few basic properties concerning the DLM units.

The fly has lost individual control over the DLM muscles which are presumably used only during flight for the generation of force to cause the wing downstroke. At rest, the motor units fire in near unison spontaneously around room temperature and below. In fact, when prostrate in cold stupor the flight motor continues to activate spontaneously with or without decapitation (12).

The DLM flight motor units start reflexly on loss of tarsal contact to initiate flight. Once on, the units are reinforced to remain on by sensory feedback. The fly can increase or decrease the rate of firing of all of the DLM flight motor units to increase or decrease the power of the wing downstroke. However, under all normal conditions, each of the DLM flight motor units almost always fire at a similar, rather low rate as a result of of presumed lateral inhibition. The basal rate of firing is 7 Hz at 20°C (12).

Preliminary studies suggest that a group of nerve cell bodies are located near the left lateral edge of the ganglion and in between the prothoracic and mesothoracic neuromeres. This group of cells includes the 5 motor neurons innervating the left DLM. Although no evidence exists, one may assume as a working hypothesis that lateral inhibitory connections between these 5 neurons occur within their immediate locality. The motor neurons send axons dorsally out the main dorsal nerve which leaves the ganglion at a point above the cell bodies. The monopolar neurons also send branches dorsally then arching toward the mid line of the ganglion to presumably connect with symmetrical branches of the right DLM motor neurons. The fine details of these latter connections are unknown, but they are presumed to support the coupling between right DLM and left DLM units.

Insect Leg Bioassays. Plots of times required to produce trains of ascending sensory impulses from topical treatment to the house fly leg (Fig. 2) or perfusion of pyrethroids through the cockroach leg (Fig. 3) were fairly similar. Compounds producing fast knockdown were more effective in producing trains of sensory pulses. Tetramethrin was slightly more potent than *cis*-methrin on the house fly leg (Fig. 2), but both tetramethrin and *cis*-methrin were far better than k-Othrin in producing ascending trains of neuron impulses. The potency of k-Othrin was more similar to that of DDT than the two pyrethroids exhibiting knockdown properties.

Flight Motor Bioassays. Flight motor pattern of a normal fly during tethered flight shows that motor units are activated at the same rate with slight differences in exact timing (Fig. 4).

Topical treatment of the house fly with 1 μg of carbofuran causes hyperactivity in a few minutes, then convulsions in about 5 minutes. By 10 minutes following treatment, the flight muscle potentials show uncoupling between the left and right DLM units (Fig. 5, traces marked: 6R & 6L). Comparison between 6L and 5L shows that the muscle potentials overlap exactly reflecting their common innervation by the same motor neuron.

This "uncoupling" between individual units of the flight motor neurons implies that carbofuran is acting on the central nervous system without an action on the peripheral nervous system. No conclusions can be drawn concerning the site or mode of action of carbofuran in causing this abnormal uncoupling response because other centrally acting neurotoxins also cause uncoupling: picrotoxin, lindane, dieldrin and organophosphates.

DDT exerts little or no direct action on the central nervous system, but can be readily characterized by monitoring flight motor potentials (12). Lethal doses of DDT cause a gradual increase in flight motor activities and splitting of flight potentials into 2 and sometimes multiple spikes (Fig. 6, arrows). This increase in activation eventually leads to a state of constant

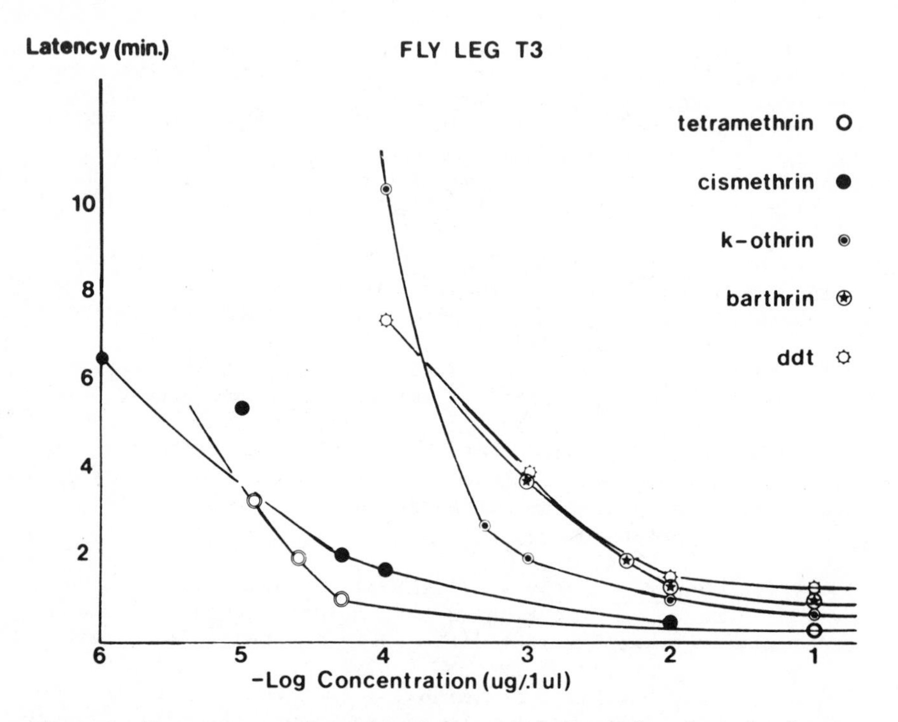

Figure 2. Dose-response relationships for insecticides applied to the isolated metathoracic leg of Musca. *Dose is expressed on the abscissa as (—)log concentration vs. latency to sensory trains in the crural nerve. Latency is short for tetramethrin and* cis-*methrin, but significantly longer for k-Othrin, which lacks activity at lower concentrations. Barthrin and DDT analogs have effects in the range of k-Othrin.*

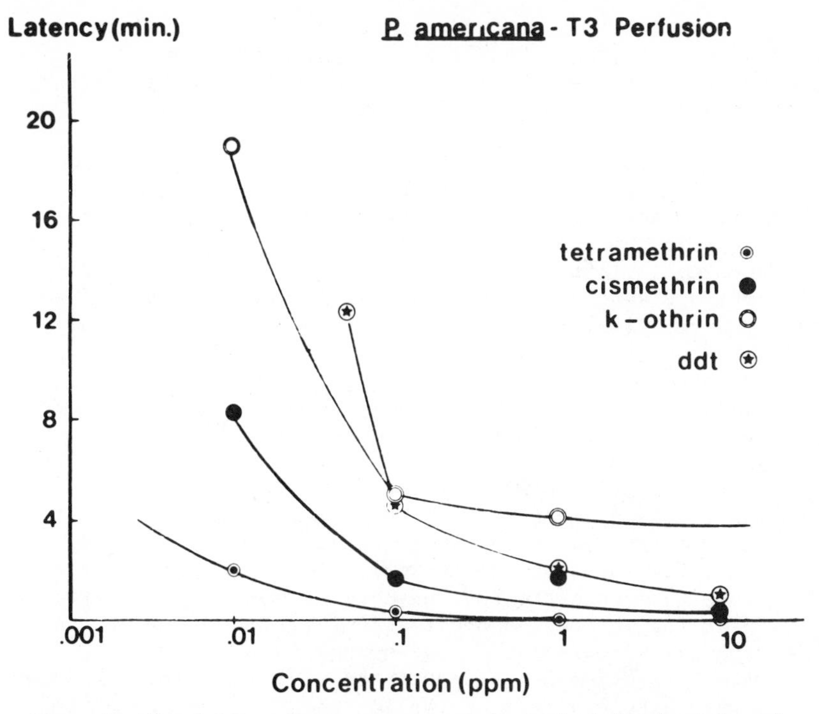

Figure 3. Plot of dose vs. latency to sensory trains in the isolated metathoracic leg of the cockroach, Periplaneta american. *The same trends apply here as in* Musca, *although* cis-*methrin assumes a more intermediate position between tetramethrin and k-Othrin.*

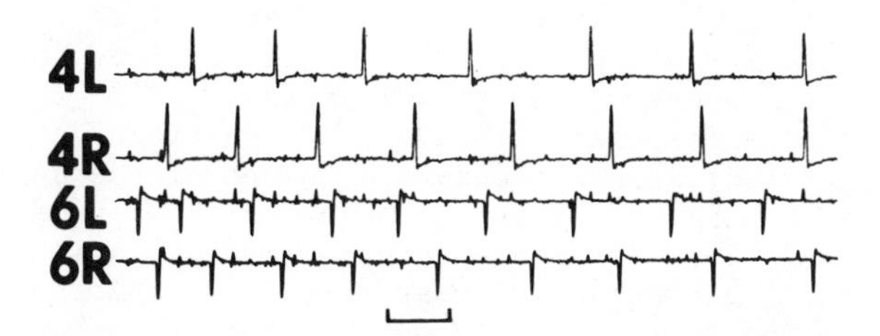

Figure 4. Flight motor pattern of a normal housefly during tethered flight. Each trace represents a discrete motor unit. A common firing frequency is maintained by all motor units, but a loose phase relationship prevents units from firing in unison. The polarity of 6R and 6L is reversed for comparison. Calibration: 100 msec.

motor output several hours after treatment (Fig. 7). Note that even in deep poisoning, the flight motor remains coupled. A brief examination of the pyrethroid, *trans*-Barthrin, on this preparation also showed a pattern of activity similar to DDT (12).

Thus, the main difference between action of DDT as a peripheral nerve poison and centrally acting insecticides remains the uncoupling caused by central poisons. Another important difference, seen from recordings of activity in the flight motor system, was in the nature of the overall activity recorded at a considerable time after poisoning. The hours following treatment by DDT analogs were characterized by a steady increase in activation in the flight motor units and, from cursory examination, in other motor units as well.

By contrast, the centrally acting insecticides, such as carbofuran, tended to produce convulsive bursts of varying length separated by relative inactivity. Often the bursts would start in flight motor units at a low rate and build gradually to higher rates with convulsive bursts of very high frequency immediately preceding complete inactivity. The period of inactivity was roughly predictable from the strength of the preceding burst.

This difference between centrally active insecticides and DDT analogs is crucial to the characterization of the action of pyrethroids for it enables one to distinguish between central and peripheral actions during poisoning.

As further evidence of central or peripheral action, pyrethroids were also examined on the leg preparations and on the isolated central nervous system to obtain some measure of inherent potency.

Tetramethrin Poisoning. When treated on the abdomen at 0.1 μg with tetramethrin, house flies developed hyperactivity very quickly. One minute following treatment, the flight muscle potentials began showing multiple discharges (Fig. 8, arrows) from single nervous impulses. The overall pattern, however, was still coupled both between units of the same muscle (Fig. 8, 6R, 4R) and between the right and left muscles (Fig. 8, cf. 6R and 4R with 6L and 4L). This can be seen from Fig. 8 where all of the units recorded are active at approximately the same times and pauses between potentials are common to all units.

At 2 minutes following tetramethrin treatment, an occasional exaggerated discharge occurred in one of the units (Fig. 9, arrow). The origin of this discharge in the nervous system has not yet been determined and is under investigation. The house fly was exhibiting hyperactivity at this point and the record obtained from the first few minutes of poisoning was very similar to records obtained 1 and 1/2 hours later. Within 4 minutes at this dose, the house flies lost locomotory ability and were considered "knocked down", but recovered over a period of several hours.

At this sublethal dose of tetramethrin there was some

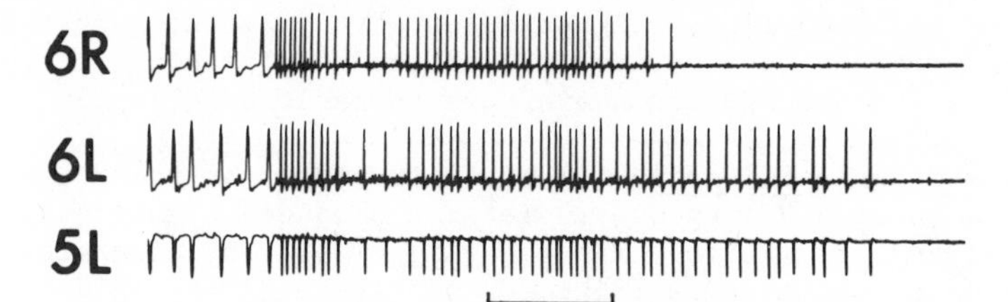

Figure 5. Disruption of coordination between motor units 6L and 6R caused by a lethal dose (1 μg) of carbofuran. This effect on central coordination is characteristic of cholinesterase poisons. The firing of 5L mirrors that of 6L, indicating common input from a single motor neuron. Calibration: 1 sec.

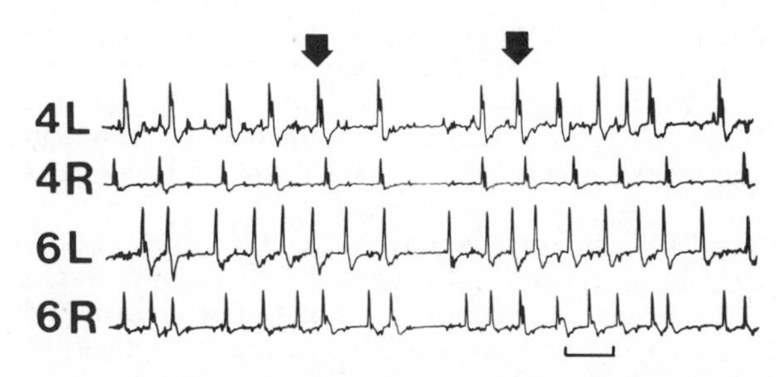

Figure 6. Flight motor pattern during tethered flight 1 hr after treatment with 1 μg DDT. Splitting of spikes is evident (arrows). Coupling between individual motor units is maintained despite symptoms of hyperactivity and locomotory in coordination at this stage of poisoning. Calibration: 100 msec.

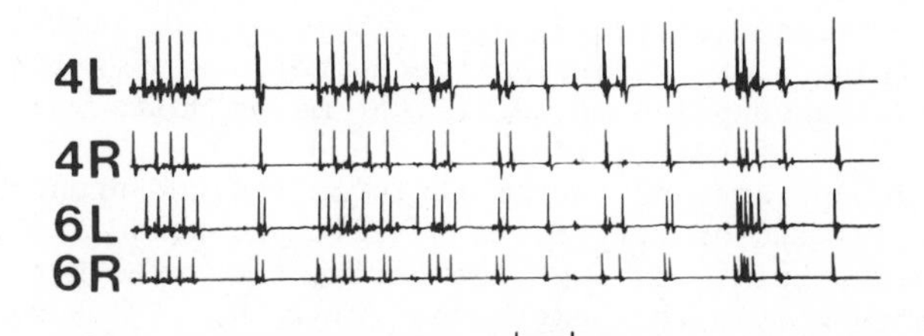

Figure 7. Condition of the flight motor almost 4 hr after treatment with 1 g DDT. The insect is in tetany and flight motor activation is almost continuous, yet the coupling between units is intact. Calibration: 0.5 sec.

evidence of uncoupling. Higher doses of tetramethrin on the abdomen produced more obvious uncoupling as did thoracic treatment (i.e., doses applied nearer the thoracic ganglion).

Cis-methrin Poisoning. Topical treatment of the house fly on the abdomen with 0.1 μg of *cis*-methrin produced highly exaggerated burst discharges within 2 minutes (Fig. 10). The discharges occurred in single units (Fig. 10, top trace A) or could occur in more than one unit (Fig. 10, all traces B). This activity was accompanied by hyperactivity.

One hour following *cis*-methrin treatment the house fly was prostrate, having lost locomotory ability in the first several minutes after treatment. There was greater evidence of uncoupling after one hour (Fig. 11, dots) while high frequency short burst discharges continued to occur in single units.

k-Othrin Poisoning. Treatment of house flies by k-Othrin at doses near the LD_{50} produced no symptoms other than an exaggerated still period (cf. 15 for description of the term still period) or quiescence which lasted for hours before the appearance of poisoning symptoms. Topical treatment of house flies by 0.1 μg of k-Othrin (25 X the LD_{50}) on the abdomen hastened the appearance of symptoms which were qualitatively similar to responses to lower doses.

14 minutes following treatment by 0.1 μg of k-Othrin exaggerated burst discharges were recorded (Fig. 12, arrows), plus some uncoupling. 20 minutes after treatment pronounced uncoupling occurred (Fig. 13). The uncoupling between 2 units in the same muscle (Fig. 13, compare 4R with 6R) was more obvious initially than uncoupling between opposing pairs of units (Fig. 13, compare 6L with its opposite unit 6R). This suggested that motor units in the same muscle were perhaps more susceptible to uncoupling than units between muscles; however, details of the uncoupling phenomenon during poisoning are being subjected to further study and will be reported on in greater detail elsewhere.

CNS Assays. The relative potency of pyrethroids on the central nervous system (CNS) of the house fly was determined. The time from treatment of the exposed CNS (thoracic ganglion in this case) by pyrethroids in various concentrations in saline to the appearance of uncoupling in the flight motor units was plotted over a range of concentrations.

The high potency of tetramethrin on the CNS preparation was surprising in view of its poorer toxicity to house fly in comparison to k-Othrin. In fact, comparing the three pyrethroids, their potency was in the same range (Fig. 14). This suggested that the pyrethroids were of similar potency in fit at the central site of action in the thoracic ganglion of the house fly and in fact, tetramethrin possessed the best fit.

Unfortunately, our sample of tetramethrin was a mixture of

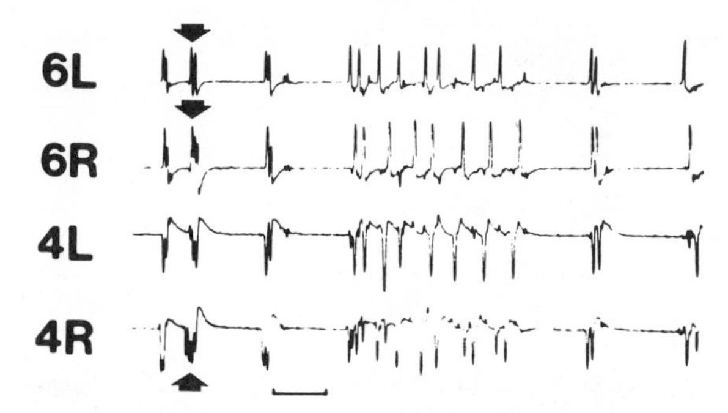

Figure 8. Flight motor output 1 min after a topical dose of 0.1 μg tetramethrin. Symptoms of hyperactivity were obvious and accompanied by slight splitting of flight motor potentials (arrows). Calibration: 100 msec.

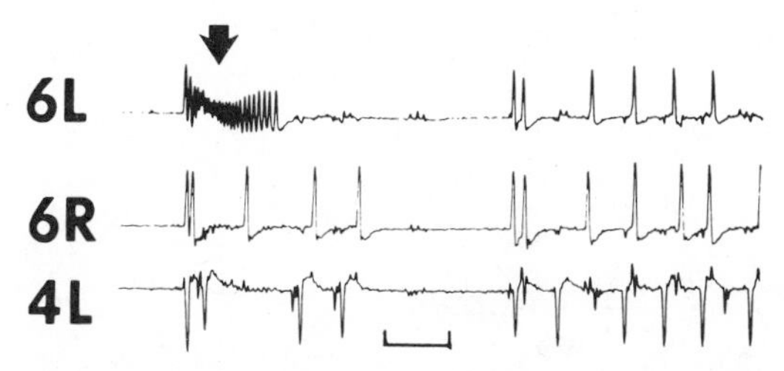

Figure 9. An exaggerated burst (arrow) in 6L occurs 2 min after treatment with 0.1 μg tetramethrin. Slight uncoupling is evident between 6L and 6R and 4R (hollow arrow). Calibration: 100 msec.

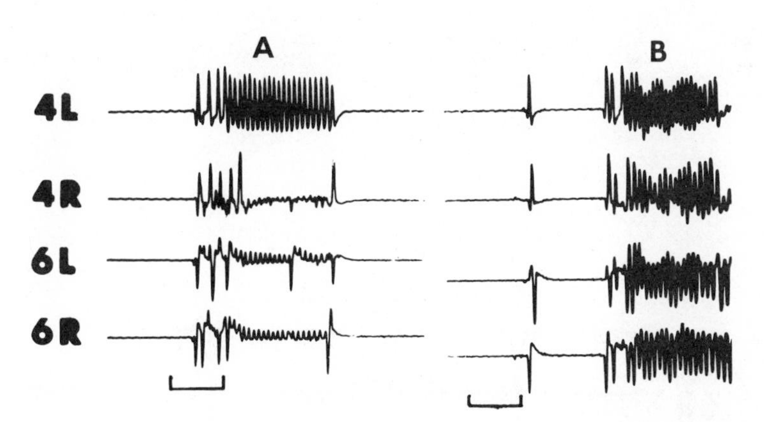

Figure 10. Topical treatment with 0.1 μg cis-methrin *elicited exaggerated burst discharging in motor units separately* (a) *or in unison* (b). *Calibration: 100 msec.*

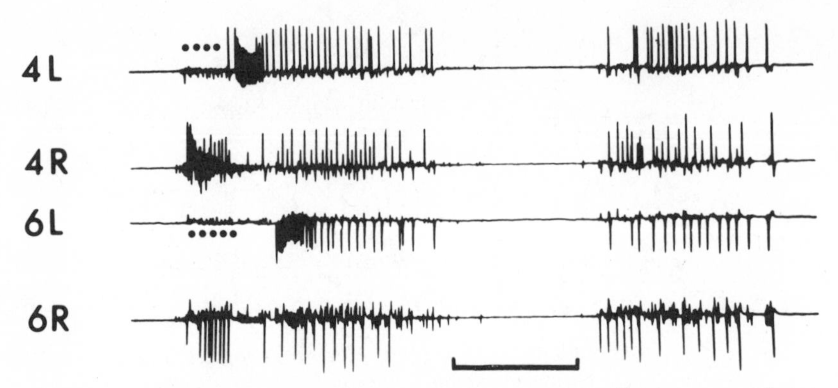

Figure 11. Flight potential pattern 1 hr after treatment with 0.1 μg cis-methrin. Flight motor is uncoupled and bursts of high frequency discharge occur separately in different motor units. Calibration: 1 sec.

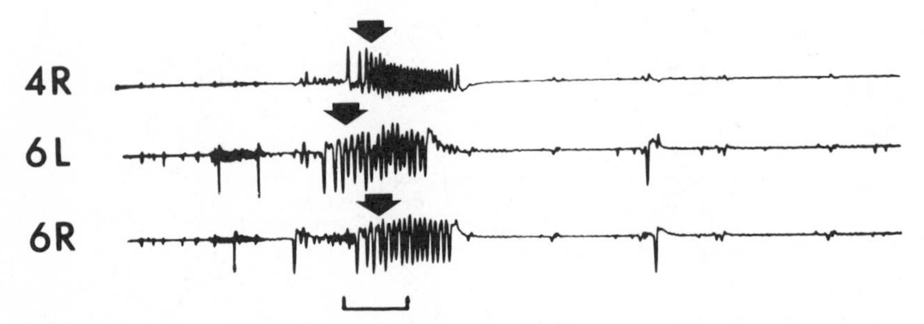

Figure 12. Onset of high frequency discharge (arrows) and uncoupling of flight motor coordination 14 min after treatment with 0.1 μg of k-Othrin ($25 \times LD_{50}$). Calibration: 100 msec.

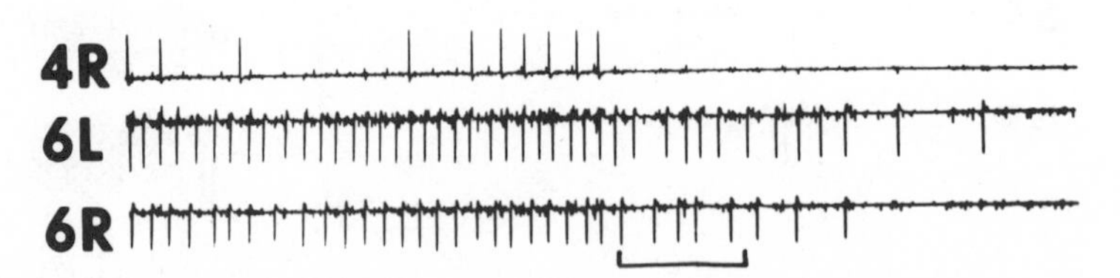

Figure 13. Pronounced uncoupling between flight motor neurons 20 min after treatment with 0.1 μg k-Othrin. Calibration: 1 sec.

cis and *trans* isomers. It would be instructive to determine the CNS activity of the resolved isomers.

The analysis of CNS activity and peripheral activity (leg assays) does provide the basis for preliminary conclusions: 1. Knockdown properties in pyrethroids are correlated with ability to produce trains of nervous impulses in peripheral sensory axons. 2. Toxicity of pyrethroids is a temporal process, i.e., if a compound can resist metabolism or detoxication for a sufficiently long time, it can accumulate at the CNS in a lethal dose regardless of knockdown properties. 3. Structure-activity studies using toxicity data would be misleading without accounting for metabolism.

These hypotheses may be examined by comparing the 3 pyrethroids examined here and Barthrin as shown in Table I.

COMPOUND	KNOCKDOWN 2 Min on Treated Paper KD Min.	PERIPHERAL Leg Trains 1.0 ng $\bar{X}$ Min.	CENTRAL Exposed CNS 10^{-7} $\underline{M}$ Uncoupling Min.	TOXICITY LD_{50} μg/fly		
				Alone	PB	SR
TETRAMETHRIN	5.75	0.5	18.7 ± 7	.29	.07	4
CIS-METHRIN	7.97	0.8	24.9 ± 7	.08	.03	2.8
k-OTHRIN	31.2	1.2	20.7 ± 5	.005	.003	1.7
BARTHRIN	-	2.3	>60	.66	.2	3.3

TABLE I. Comparative knockdown, peripheral sensory action, central nervous action and toxicity for 4 pyrethroids.

k-Othrin at 1.7 has the lowest synergistic ratio with piperonyl butoxide of the compounds compared in Table I. Tetramethrin with a synergistic ratio of 4 and Barthrin at 3.3 are higher than k-Othrin suggesting the latter are detoxified more readily. Even higher synergistic ratios are reported for tetramethrin with other synergists (25, 26).

Barthrin has poor CNS potency which could account for its action resembling that of DDT more than that of centrally active pyrethroids. As a coincidence, the peripheral actions of DDT and Barthrin (Fig. 2) are very similar in bioassay on the house fly leg. These factors suggest that pyrethroids can be toxic by virtue of a DDT-like action (Barthrin), but that the modern synthetic pyrethroids as pioneered by Dr. Elliott owe their potency to an improved action of the central nervous system. In effect, there are 2 sites of action and the peripheral responses mask the more important effects in the central nervous system.

Although there is a correlation between knockdown and the ability to produce trains of sensory nerve impulses for the 4

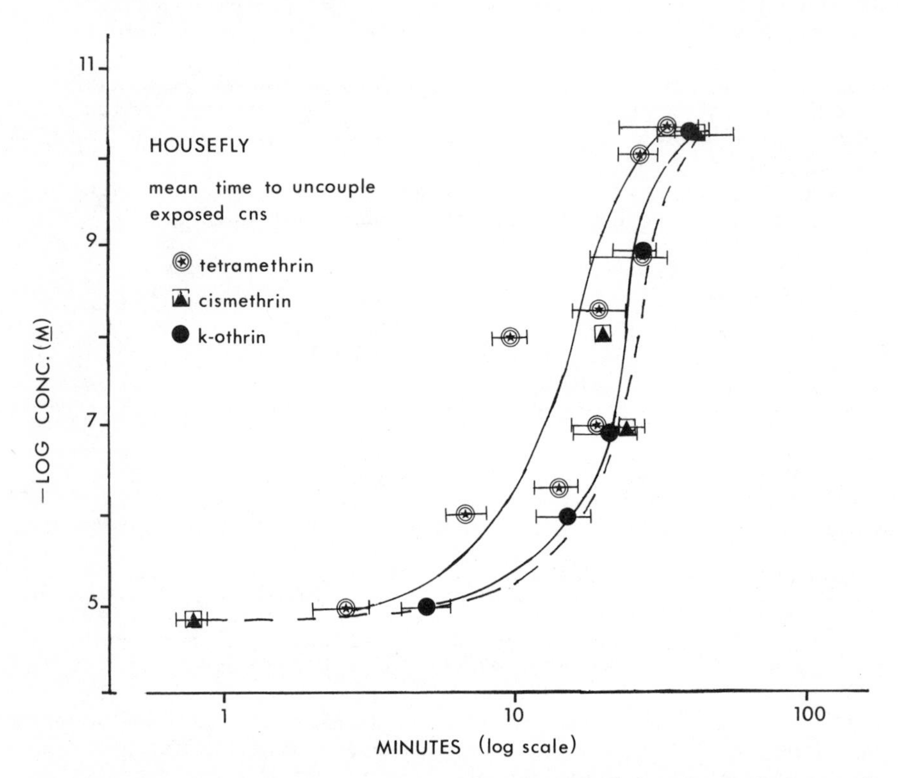

Figure 14. Dose-response curves for pyrethroids applied directly on the exposed CNS in saline. Dose is plotted on ordinate as (−) log concentration in mol/L against mean time to uncoupling on the abscissa. Tetramethrin shows slightly higher potency than cis-methrin *or k-Othrin, but all fall in approximately the same range.*

pyrethroids examined, this does not prove that knockdown is caused entirely by peripheral actions. Uncoupling indicates chemical poisoning in the central nervous system. However, we do not know yet if central nervous poisoning can occur without uncoupling or before uncoupling.

Therefore, the friendly disagreement between Paul Burt and ourselves concerning central versus peripheral poisoning will not be resolved until more is known about the central site and mode of action of pyrethroids. There is little doubt, however, that some pyrethroids are acting both on peripheral nerves and on the CNS. At present, it is not entirely possible to assess the contribution from peripheral action and that from central action to symptoms of poisoning.

Acknowledgement

The authors wish to thank Drs. Richard J. Hart and Charles Potter of Wellcome Research Laboratories, Berkhamsted, England for providing compounds and for their very useful discussions. Some of the results included here are from work by V. Salgado and J. Kennedy to be published in greater detail elsewhere.

The research was supported in part by Environmental Protection Agency Grant No. R-804345-01. The contents do not necessarily reflect the views and policies of the Environmental Protection Agency, nor does mention of trade names or commercial products constitute endorsement or recommendation for use.

Literature Cited

1. Vinson, E. B. and C. W. Kearns. (1952). Temperature and the action of DDT on the American roach. *J. Econ. Ent.* 45: 484.
2. Blum, M. S. and C. W. Kearns. (1956). Temperature and the Action of Pyrethrum in the American Cockroach. *J. Econ. Ent.* 49: 862.
3. Narahashi, T. (1971). Effects of insecticides on excitable tissues. *Adv. Insect Physiol.* 8: 1.
4. Singh, K. M., S. Pradhan and C. Dakshinamurti. (1972). Differential susceptibility of insect neuron to insecticides. *Indian J. Ent.* 34(4): 263-271.
5. Roeder, K. D. and E. A. Weiant. (1946). The site of action of DDT in the cockroach. *Science N.Y.* 103: 304-306.
6. Roeder, K. D. and E. A. Weiant. (1948). The effect of DDT on sensory and motor structure of the cockroach leg. *J. Cell Comp. Physiol.* 32: 175-186.
7. van den Bercken, J., L. M. A. Akkermans and J. M. van der Zalm. (1973). DDT-like action of allethrin in the sensory nervous system of *Xenopus laevis*. *Europ. J. Pharmacol.* 21: 95-106.

8. Burt, P. E. and R. E. Goodchild. (1974). Knockdown by pyrethroids: its role in the intoxication process. *Pestic. Sci.* 5: 625-633.
9. Page, A. B. P. and R. E. Blackith. (1949). The mode of action of pyrethrum synergists. *Ann. appl. Biol.* 36: 244-249.
10. Burt, P. E. and R. E. Goodchild. (1971). The site of action of pyrethrin I in the nervous system of the cockroach, *Periplaneta americana*. *Ent. exp. & Appl.* 14: 179-189.
11. Miller, T., L. J. Bruner and T. R. Fukuto. (1971). The effect of light, temperature, and DDT poisoning on housefly locomotion and flight muscle activity. *Pestic. Biochem. Physiol.* 1: 483.
12. Miller, T. and J. M. Kennedy. (1972). Flight motor activity of house flies as affected by temperature and insecticides. *Pestic. Biochem. Physiol.* 2: 206.
13. Miller, T. and J. M. Kennedy. (1973). *In vivo* measurement of house fly temperature, flight muscle potentials, heartbeat and locomotion during insecticide poisoning. *Pestic. Biochem. Physiol.* 3: 370.
14. Miller, T., J. M. Kennedy, C. Collins, and T. R. Fukuto. (1973). An examination of temporal differences in the action of carbamate and organophosphorus insecticides on house flies. *Pestic. Biochem. Physiol.* 3: 447.
15. Miller, T. A. (1976). Distinguishing between carbamate and organophosphate insecticide poisoning in house flies by symptomology. *Pestic. Biochem. Physiol.* 6: 307-319.
16. Burt, P. E. (1974). Personal communication.
17. Hart, R. J. (1975). Personal communication.
18. Wilkens, L. A. and G. E. Wolfe. (1974). A new electrode design for *en passant* recording, stimulation, and intracellular dye infusion. *Comp. Biochem. Physiol.* 48A: 217-220.
19. Miller, T. and J. James. (1976). Chemical sensitivity of the hyperneural nerve-muscle preparation of the American cockroach. *J. Insect Physiol.* 22: 981-988.
20. Nachtigall, W. and D. M. Wilson. (1967). Neuromuscular control of dipteran flight. *J. Exp. Biol.* 47: 77.
21. Berridge, M. J. (1966). Metabolic pathways of isolated Malphigian tubules of the blowfly functioning in an artificial medium. *J. Insect Physiol.* 12: 1523-1538.
22. Thomson, A. J. (1975). Regulation of crop contraction in the blowfly, *Phormia regina* Meigen. *Can. J. Zool.* 53: 451-455.
23. Mulloney, B. (1970b). Organization of flight motor neurons in Diptera. *J. Neurophysiol.* 33: 86-95.
24. Wilson, D. M. (1968). The nervous control of insect flight and related behavior. *Adv. Insect Physiol.* 5: 289.

25. Jao, L. T. and J. E. Casida. (1974). Esterase inhibitors as synergists for (+)-*trans*-Chrysanthemate insecticide chemicals. *Pestic. Biochem. Physiol.* 4: 456-464.
26. Miyamoto, J. and T. Suzuki. (1973). Metabolism of tetramethrin in houseflies *in vivo*. *Pestic. Biochem. Physiol.* 3: 30.

10

Synthetic Route to the Acid Portion of Permethrin

M. S. GLENN and W. G. SCHARPF

FMC Corp., Agricultural Chemical Division, Box 8, Princeton, N.J. 08540

The diene was originally prepared by Farkas, Kourim, and Sorm (1) by the following synthetic route:

Cl_3CCHO + [isobutylene] $\xrightarrow{AlCl_3}$ Cl_3C–CH(OH)–CH$_2$–C(=CH$_2$)CH$_3$ 85%

$\xrightarrow[pyridine]{Ac_2O}$ Cl_3C–CH(OAc)–CH$_2$–C(=CH$_2$)CH$_3$ 89% $\xrightarrow[HOAc]{Zn}$ Cl_2C=CH–CH$_2$–C(=CH$_2$)CH$_3$ 62%

$\xrightarrow{pTSA}$ Cl_2C=CH–CH=C(CH$_3$)$_2$ 81%

Although the Prins reaction proceeded in good yield with only a small amount of polymer formation, acetylation and reduction required large amounts of pyridine and zinc which were too costly for commercialization. The overall yield of 1,1-dichloro-4-methyl-1,3-pentadiene was only 38%.

Our reaction sequence consisted of three steps with an overall yield of 57%, although yields were not maximized. Only low cost, commercially-available chemicals were used.

The Darzens-Kondakov reaction of 1,1-dichloroethene with isobutyryl chloride proceeded as follows:

$$Cl_2C{=}CH_2 + ClCCH(CH_3)_2 \xrightarrow{AlCl_3}$$

(with C=O on the acyl carbon)

Cl_3C–CH$_2$–C(=O)–CH(CH$_3$)$_2$ $\xrightarrow[\text{or base}]{\text{heat}}$ Cl_2C=CH–C(=O)–CH(CH$_3$)$_2$

The reaction was developed by Heilbron, Jones, and Julia (2); improved by Soulen et al. (3) who used carbon tetrachloride as a diluent; and finally by Atvin, Levkovskaya, and Mirskova (4) who used potassium carbonate to remove a mole of hydrogen chloride. The thermal removal of hydrogen chloride was difficult for this compound and required steam distillation followed by fractional distillation. We found that removal of hydrogen chloride was best achieved by using sodium carbonate. The reaction proceeded smoothly at 16 gram moles and gave a minimum of 67% distilled yield. Other Lewis acids such as stannic chloride and ferric chloride gave zero or a poor yield respectively.

The reduction of the ketone to the alcohol was attempted by the catalytic method of Adams (5) (platinum oxide and ferrous chloride), but selective reduction was not obtained.

Cl_2C (ketone, O) $\xrightarrow[\text{isopropoxide}]{\text{aluminum}}$ Cl_2C (alcohol, OH)

A standard reduction (6) using an equimolar amount of commercial aluminum isopropoxide gave an 82% yield of the alcohol. When one-quarter mole of freshly prepared aluminum isopropoxide was used per mole of ketone, a 62% yield of the alcohol was isolated. Equimolar amounts of freshly prepared isopropoxide afforded a 94% yield of the distilled alcohol. The Prins reaction of isobutyraldehyde and 1,1-dichloroethene did not give the desired alcohol because of aldol formation.

The dehydration of the alcohol gave 1,1-dichloro-4-methyl-1,3-pentadiene and smaller amounts of the corresponding 1,4-diene. The results from various acidic catalyst were shown in the following table.

Cl_2C (OH) $\xrightarrow{H^+}$ Cl_2C 1) + Cl_2C 2)

H^+	1)	2)
Acid Clay (Superfiltrol)	89%	0.5%
$K_2S_2O_7$*	26	-
pTSA	20	-
$KHSO_4$	14	-
H_3PO	57	9
Superfiltrol*	83	0.5

*Azeotropic removal

The best commercial method used 1.0-1.5% of acid clay at 100-120° and an inert gas to aid in the removal of water. After the evolution of water was completed the diene was distilled directly from the reaction flask. Since the 1,4-diene may be isomerized to the 1,3-diene with p-toluenesulfonic acid, a total yield of 90% was obtained. The acid clay, Superfiltrol, was a low volatile material obtained from the Filtrol Company, Los Angeles, California.

The reaction of the diene with ethyl diazoacetate as described by Farkas (1) was repeated to give a 37% conversion and a 71% yield of ethyl 3-(2,2-dichloro-vinyl)-2,2-dimethylcyclopropanecarboxylate.

The cost evaluation for 1,1-dichloro-4-methyl-1,3-pentadiene gave a unit cost of $1.87 per pound based on the price of raw materials as of October 1975.

The commercialization of the preparation of ethyl diazoacetate and its reaction to form the ethyl ester of the permethrin acid would be similar to that of past allethrin synthesis (7).

Literature Cited

1. J. Farkas, P. Kourim, and F. Sorm, Coll. Czech. Chem. Comm., 24, 2230 (1959).
2. I. Heilbron, E. R. Jones, and M. Julia, J. Chem. Soc., 1949, 1430.
3. R. L. Soulen, D. G. Kundiger, S. Searles and R. A. Sanchez, J. Org. Chem., 32, 2661 (1967).
4. A. S. Atavin, G. G. Levkovskaya and A. N. Mirskova, J. Org. Chem. (USSR), 9, 318 (1973).
5. R. Adams, J. Amer. Chem. Soc., 47, 3064 (1925).
6. A. L. Wilds, Organic Reactions, II, 178 (1944).
7. H. J. Sanders and A. W. Taft, Industrial and Engineering Chem., 46, 414 (1954).

11

Novel Routes to 1,1-Dichloro-4-methyl-1,4-pentadiene and 1,1-Dichloro-4-methyl-1,3-pentadiene

MANUEL ALVAREZ and MORRIS L. FISHMAN

FMC Corp., Agricultural Chemical Division, Box 8, Princeton, N.J. 08540

Pyrethroids, in general, have a high degree of activity as insecticides while showing a low mammalian toxicity. The natural pyrethroids cannot be used commercially to protect agricultural crops mainly because of their high cost and their poor photostability. The synthesis of 3-phenoxybenzyl-3-(2,2-dichlorovinyl)-2,2-dimethylcyclopropanecarboxylate (NRDC 143) has been reported (1). This material has high insecticidal activity and low mammalian toxicity. It also has greater photostability than the natural pyrethroids.

Cl_2C ... CO_2CH_2 ...

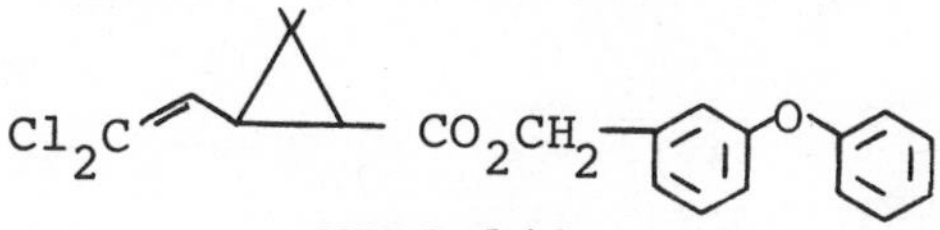

NRDC 143

NRDC 143 has been prepared from its corresponding ethyl ester. Acid hydrolysis of ethyl 3-(2,2-dichlorovinyl)-2,2-dimethylcyclopropanecarboxylate 1 formed 3-(2,2-dichlorovinyl)-2,2-dimethylcyclopropane carboxylic acid 2. Treatment of 2 with thionyl chloride gave the corresponding acid chloride 3 in an 80% overall yield from 1. Treatment of 3 with 3-phenoxybenzyl alcohol formed the desired NRDC 143.

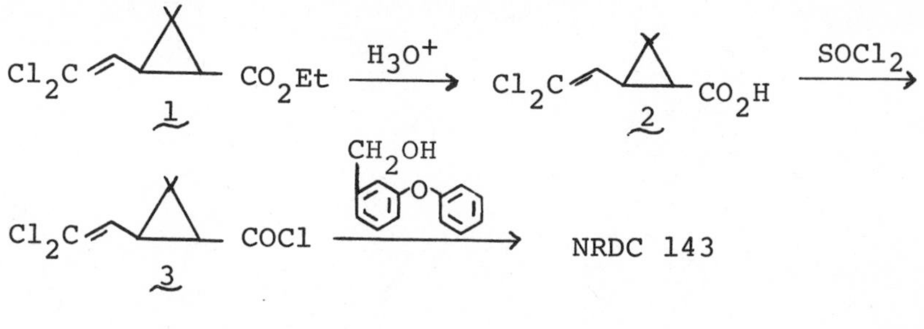

Cl_2C ... CO_2Et (1) $\xrightarrow{H_3O^+}$ Cl_2C ... CO_2H (2) $\xrightarrow{SOCl_2}$

Cl_2C ... $COCl$ (3) $\xrightarrow{CH_2OH \text{ (3-phenoxybenzyl alcohol)}}$ NRDC 143

$Cl_3C\text{-}CH(OH)\text{-}CH_2\text{-}C(=CH_2)CH_3$

4

$2e^-$ $-Cl^-$

k_1 e^-, $-Cl^-$

$Cl_2\dot{C}\text{-}CH(OH)\text{-}CH_2\text{-}C(=CH_2)CH_3$

k_3

$Cl_2C\text{-}CH(OH)\text{-}CH_2\text{-}C(=CH_2)CH_3$
|
$Cl_2C\text{-}CH(OH)\text{-}CH_2\text{-}C(=CH_2)CH_3$

k_2 e^-

$Cl_2C^{(-)}\text{-}CH(OH)\text{-}CH_2\text{-}C(=CH_2)CH_3$

k_4 proton source

$Cl_2CH\text{-}CH(OH)\text{-}CH_2\text{-}C(=CH_2)CH_3$

8

k_5

$Cl_2C{=}CH\text{-}CH_2\text{-}C(=CH_2)CH_3$

6

Figure 1

Compound 1 has been prepared by the method of Farkas (2). Condensation of chloral with isobutylene gave 1,1,1-trichloro-2-hydroxy-4-methyl-4-pentene 4 and its isomeric 3-pentene 13. Acetylation of 4 and 13 with a mixture of acetic anhydride and pyridine gave 2-acetoxy-1,1,1-trichloro-4-methyl-4-pentene 5 and its corresponding isomer, 14. Treatment of the acetoxy mixture with zinc-acetic acid gave the expected 1,1-dichloro-4-methyl-1,4-pentadiene 6 and the 1,3-pentadiene 7. Isomerization of 6 and 7 with p-toluenesulfonic acid gave the desired 1,1-dichloro-4-methyl-1,3-pentadiene 7, which upon treatment with ethyl diazoacetate gave the ethyl cyclopropanecarboxylate 1.

Cl_3CCHO + $(CH_3)_2C{=}CH_2$ $\xrightarrow{AlCl_3}$ Cl_3C (OH) 4 +

Cl_3C (OH) 13 $\xrightarrow[\text{pyridine}]{Ac_2O}$ Cl_3C (OAc) 5 + Cl_3C (OAc) 14 $\xrightarrow[\text{HOAc}]{Zn}$

Cl_2C 6 + Cl_2C 7 $\xrightarrow{pTSA}$ 7 $\xrightarrow{N_2CHCO_2Et}$ 1

In order to avoid the zinc reaction, routes more amenable to commercialization were studied. Some electrochemical reductive eliminations of 4 and other 2-substituted analogs of 4 to intermediate 6 were studied in our laboratories.

Some electrochemical pathways for the electrochemical reductive elimination of 4 to the desired diene 6 are shown in Figure 1. Compound 4 could be reduced by a one electron step to first form the dichloro radical which could either dimerize at that point or acquire another electron to form the carbanion. The carbanion could then be protonated by a proton source, such as a protolytic solvent, to form 1,1-dichloro-2-hydroxy-4-methyl-4-pentene 8 or undergo elimination to form directly the desired diene 6. As expected for organo chlorine compounds, the rate of two electron addition was rapid enough that radical formation was not significant and the observed products were consistent with the formation of a carbanion intermediate.

A divided electrolysis cell, as diagrammed in Figure 2, was used in the electrochemical reactions so that oxidation of the starting material or product could not occur at the anode. The cathode and anode are generally separated by means of a permeable barrier such as fritted discs or ion exchange membranes. Three

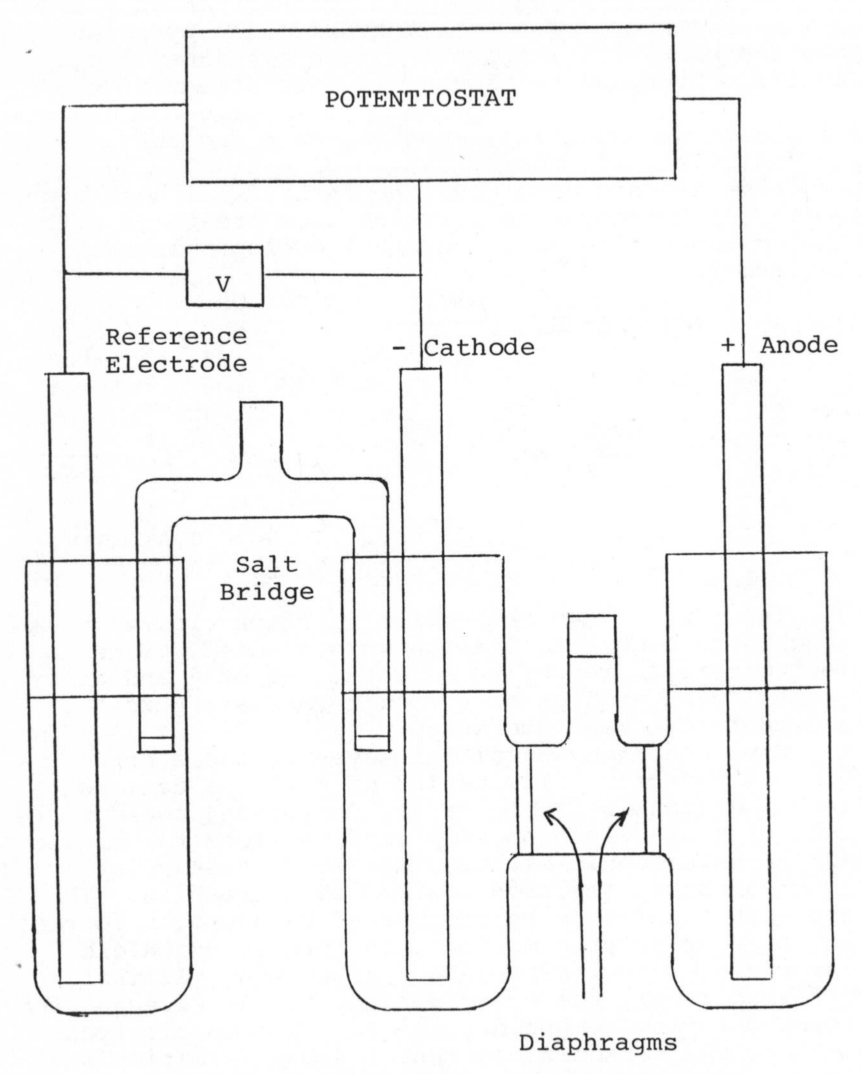

Figure 2. Schematic representation of electrolysis experiment

electrodes were used in each experiment; a cathode, an anode and a reference electrode such as the saturated calomel electrode. By using the reference electrode, the cathode potential can be easily controlled at a pre-set value relative to a reference electrode which is situated as near to the cathode surface as is experimentally possible. A center compartment (salt bridge) filled with a catholyte (fluoroborates) greatly minimizes intercompartment diffusion of catholyte and anolyte. Dilute sulfuric acid is a convenient anolyte and the overall anode reaction is electrolysis of water to oxygen and protons. A photograph of one electrolysis cell used in our laboratories is shown in Figure 3.

Electrolysis of 4 in dimethylformamide, using a mercury cathode, gave a product which consisted of 35.2% of 6, according to vapor-phase chromatographic analysis. The mass spectra of 6 was identical to that of 6 prepared via the Farkas route. A double salt bridge was used in this experiment to reduce the leakage of proton sources from the reference electrode. Dimethylformamide was used as the solvent in this reaction mainly because it is an aprotic solvent which coordinates with the OH group to favor reductive elimination of the OH group. However, there was still sufficient proton activity from 4 to form 8.

OH

Cl_3C (4) $\xrightarrow[\text{Hg-DMF}]{2e^-}$ Cl_2C (6) + Cl_2HC (8)

OH

The electroreduction of compounds with better leaving groups than hydroxy and also free of labile protons were studied. Electrolysis of 5 in acetonitrile using a mercury cathode gave a product which contained 41% of the desired diene 6 according to gc.

OAc

Cl_2C (5) $\xrightarrow[\text{Hg-}CH_3CN]{2e^-}$ Cl_2C (6)

In the search for an even better leaving group, we chose to make the methanesulfonate. Treatment of 4 with methanesulfonyl chloride gave 1,1,1-trichloro-4-methyl-4-penten-2-yl methanesulfonate 9, mp 60-61°C, in a 69% yield: nmr ($CDCl_3$), δtms 3.2 (s, 3H, $-OSO_2CH_3$), no evidence of hydroxyl protons; ms, 185 ($M-CH_3SO_2$, 3Cl present), 149 (185-HCl, 2Cl present), molecular ion (280) observed only with chemical ionization; Anal. Calc. for $C_7H_{11}Cl_3O_3S$: C, 29.84; H, 3.91; Cl, 37.83; S, 11.37. Found: C, 29.89; H, 4.06; Cl,

37.76; S, 11.48.

Electrochemical reductive elimination of 9 in acetonitrile using a platinum cathode gave the desired diene 6 in a nearly quantitative yield. The current efficiency in this case was essentially quantitative.

$$Cl_3C\text{-}CH(OH)\text{-}CH_2\text{-}C(CH_3)=CH_2\ (\mathbf{4}) \xrightarrow{CH_3SO_2Cl} Cl_3C\text{-}CH(OSO_2CH_3)\text{-}CH_2\text{-}C(CH_3)=CH_2\ (\mathbf{9}) \xrightarrow[Pt\text{-}CH_3CN]{2e^-} Cl_2C=CH\text{-}CH_2\text{-}C(CH_3)=CH_2\ (\mathbf{6})$$

In some instances, synthesized samples of the above methane sulfonate decomposed after storage giving black, tarry and acidic products. The following mechanism is suggested for this observed decomposition

$$Cl_3C\text{-}CH(OSO_2CH_3)\text{-}CH(H)\text{-}C(CH_3)=CH_2\ (\mathbf{9}) \longrightarrow Cl_3C\text{-}CH=CH\text{-}C(CH_3)=CH_2 + CH_3SO_3H$$

↓ polymeric material

Since the methane sulfonate 9 may be unstable, a compound with a better leaving group than an acetoxy one but not as good as a mesyloxy one was prepared. Treatment of 4 with thionyl chloride gave 1,1,1-trichloro-4-methyl-4-penten-2-yl chlorosulfinate 10, which could not be isolated in pure form by normal distillation. The boiling point of 10 was 69-82° (0.09-0.1 mm) (62% purity by vapor phase chromatographic analysis). GC-MS analysis of the distilled product established that the main component gave m/e 284 (M), 248 (M-HCl) and 185 ($M-SO_2Cl$, 3 Cl present). Electrochemical treatment of 10 in acetonitrile using a platinum cathode gave a product which contained 10% of the desired diene 6 according to gc.

$$Cl_3C\text{-}CH(OH)\text{-}CH_2\text{-}C(CH_3)=CH_2\ (\mathbf{4}) \xrightarrow{SOCl_2} Cl_3C\text{-}CH(OSOCl)\text{-}CH_2\text{-}C(CH_3)=CH_2\ (\mathbf{10}) \xrightarrow[Pt\text{-}CH_3CN]{2e^-} Cl_2C=CH\text{-}CH_2\text{-}C(CH_3)=CH_2\ (\mathbf{6})$$

Treatment of 4 with phosphorus trichloride gave 1,1,1,4-tetrachloro-2-hydroxy-4-methylpentane 11, mp 79-81°C in a 25% yield, nmr ($CDCl_3$), δtms 4.4 (m, 1H, -C<u>H</u>OH), δtms 2.25 (m, 2H, $-CH_2-$), δtms 1.7 (s, 6H, $-C(CH_3)_2$); Anal. Calc for $C_6H_{10}Cl_4O$: C, 30.00, H, 4.17; Cl: 59.17. Found: C, 29.89; H, 4.27; Cl, 59.09.

Figure 3

Figure 4

Reaction of 4 with phosphorus pentachloride gave 1,1,1-trichloro-4-methyl-2,4-pentadiene 12, MS, 185 (MH^+, 3Cl present), 149 (MH^+-HCl).

Cl_3C 4 → (PCl_3) Cl_3C 11 (OH, Cl); → (PCl_5) Cl_3C 12

As shown previously, condensation of chloral with isobutylene gave a mixture of 4 and 13. Acetylation of 13 with a mixture of acetic anhydride and pyridine gave 2-acetoxy-1,1,1-trichloro-4-methyl-3-pentene 14, bp 85-90°C (4-4.3 mm) (2) in an 82% yield. Electrochemical reductive elimination of 14 in acetonitrile using a mercury cathode gave the expected diene 7 in a 48% yield based on gc area %.

Cl_3C (OH) 13 → (Ac_2O, pyridine) Cl_3C (OAc) 14 → ($2e^-$, Hg-CH_3CN) Cl_3C 7

The electrochemical reactions can be run in a large divided electrolysis cell, shown in Figure 4, which is similar in design to a commercial unit. Flow cells can also be used in these electrochemical experiments.

Acknowledgments

The authors wish to thank Messrs. Ken Goldsmith and Harold Jarrow for their technical assistance, Mr. Robert Rosen for help in interpreting mass spectra and Mr. Robert Schipmann for construction of the electrolysis cells.

Literature Cited

1. Burt, P. E., Elliott, M., Farnham, A. W., Janes, N. F., Needham, P. H. and Pulman, D. A., Pestic Sci. (1974), 5, 791-799.
2. Farkas, J., Kourim, P. and Sorm, F., Collection Czechoslov. Chem. Commun. (1959), 24, 2230-2236.

12

New Synthesis of the Acid Moiety of Pyrethroids

KIYOSI KONDO, KIYOHIDE MATSUI, and AKIRA NEGISHI

Sagami Chemical Research Center, 4-4-1 Nishi-Ohnuma,
Sagamihara, Kanagawa 229 Japan

Since the discovery of permethrin (1) and its bromo analogue (2), interest in the structural modification of natural pyrethroids has been renewed owing to their potential use as agricultural pesticides as well as household insecticides. The reaction of ethyl diazoacetate with 1,1-dichloro-4-methyl-1,3-pentadiene was originally used by Farkas (3) in the synthesis of the acid moiety of permethrin. Most of the acid moieties, however, were usually prepared by the ozonolysis of the parent chrysanthemate followed by condensation of the resulting 2-formyl-3,3-dimethylcyclopropanecarboxylate (caronaldehyde) with appropriate Wittig reagents (4,5).

Cl, Cl —(N_2CHCO_2R)→ R, O, O, Cl, Cl ←($Ph_3P{=}CCl_2$)— R, O, O, CHO

We have developed a new and generally applicable method for the preparation of these potentially useful synthetic pyrethroids. The method is based on the reaction between allylic alcohol and orthoester to produce γ-unsaturated carboxylate, followed by the addition of carbon tetrahalide to the double bond, or allylic bromination with N-bromosuccinimide. The dehydrolhalogenation of the resulting halides afforded the desired cyclopropanecarboxylates.

Synthesis of the Dihalovinyl Analogues of Chrysanthemate

The condensation of 3-methyl-2-butenol (I) with

triethyl orthoacetate (IIa) in the presence of acid catalyst was performed at 140-160° according to the method described by Johnson (6) to give ethyl 3,3-dimethyl-4-pentenoate (IIIa) (7). The α-substituted analogue of III can similarly be prepared by the condensation of I with appropriate orthocarboxylates (8,9).

I (OH) + $R^1CH_2C(OEt)_3$ II —cat.→ III

II a: R^1=H
b: R^1=Me

III a: R^1=H
b: R^2=Me

III —CX_4/cat.→ IV

IV —base→ V

V a: R^1=H, X=Cl
b: R^1=H, X=Br
c: R^1=Me, X=Cl

IV a: R^1=H, X=X'=Cl
b: R^1=H, X'=Br, X=Cl
c: R^1=H, X=X'=Br
d: R^1=Me, X=X'=Cl

The effect of the variation of reaction condition on the yield of III is summarized in Table I.

Table I. Preparation of 4-Pentenoates

R^1	mol.ratio of I:II	Catalyst*	Time (hr)	Yield of III
H	1:2	phenol	25	76
H	1:1.05	phenol	25	60
H	1:2	H_3PO_4	6	81
H	1:3	H_3PO_4	4	93
H	1:2	oxalic acid	27	65
H	1:2	i-butyric acid	23	70
H	1:2	$Hg(OAc)_2$	23	69
H	1:2	hydroquinone	23	51
Me	1:2	phenol	24	70
Et**	1:1.5	phenol	25	57

* The amount of catalyst was usually 1-5 mole % based on I.

** Trimethyl orthobutyrate was used.

When the reaction was started from 1:1 molar mixture of I and IIa, there was observed the formation of 3-methyl-2-butenyl 3,3-dimethyl-4-pentenoate as by-product in ca. 20% yield. Thus, in order to attain the maximum yield of III, an excess amount of orthoacetate must be used. Most of the excess orthoester can, however, be recovered by fractional distillation. Among the catalysts examined, phosphoric acid seems to be the best in view of the yield and reaction rate.

The addition of carbon tetrahalide to the 4-pentenoate III was easily achieved either in the presence of radical initiator (10) or by irradiation (11). Thus, heating of a solution of IIIa and benzoylperoxide (BPO) in carbon tetrachloride for 20 hrs at 80° afforded ethyl 3,3-dimethyl-4,6,6,6-tetrachlorohexanoate (IVa), b.p. 107-108°/0.3 mm., in 86% yield. When the ester IIIa was treated with bromotrichlormethane in the presence of azobisisobutyronitrile (AIBN), ethyl 3,3-dimethyl-4-bromo-6,6,6-trichlorohexanoate (IVb), b.p. 102-105°/0.1 mm., was obtained in 89% yield. Similarly, radical addition of carbon tetrabromide to IIIa by irradiation with visible light produced the corresponding tetrabromo analogue IVc, b.p. 144°/0.2 mm., in 60% yield. Other initiators, such as transition metal-amine complexes (12,13,14), which are well-known as effective catalysts for the addition of polyhaloalkanes to olefins, can also be used. Some typical examples including the results with α-substituted analogues are collected in Table II.

Table II. Addition of CX_4 to 4-Pentenoates

R^1	CX_4	Catalyst	Temp.	Time	Yield of IV
H	CCl_4	BPO	80	20	86
H	CCl_4	$FeCl_3 6H_2O$-$BuNH_2$	120	20	87
H	CCl_4	$Cu(OAc)_2$-$BuNH_2$	90	20	87
H	CCl_4	Cu_2O-$BuNH_2$	90	20	85
H	$CBrCl_3$	AIBN	100	10	89
H	$CBrCl_3$	BPO	80	20	87
H	$CBrCl_3$	$FeCl_3 6H_2O$-$BuNH_2$	120	15	52
H	CBr_4	AIBN	120∿130	5	45
H	CBr_4	hν	r.t.	10	60
Me	CCl_4	BPO	130∿140	20	70
Me	$CBrCl_3$	BPO	100	10	81
Me	CCl_4	$FeCl_3 6H_2O$-$BuNH_2$	120	10	49
Et*	CCl_4	$FeCl_3 6H_2O$-$BuNH_2$	120	10	80

* Me-ester

Treatment of the ester IV with two molar equivalents of base induced cyclization and dehydrohalogenation simultaneously to afford dihalovinylcyclopropanecarboxylate V. The cis:trans ratio of the resulting ester V varied depending on the reaction conditions used. Typical examples are shown in Table III.

Table III. Dihalovinylcyclopropanecarboxylates V

Starting halide	Reaction Conditions base	solvent	temp.	time	Product V	cis/trans	yield (%)
IVb	t-BuOK	THF	60°	4	Va	45/55	70
IVa	t-BuOK	THF	r.t. 60°	3 3.5	Va	50/50	73
IVa	t-BuONa	THF	5°	3	Va	50/50	92
IVa	NaOEt	EtOH	r.t. 80°	2 1.5	Va	34/66	94
IVa	KOEt	EtOH	r.t. 80°	2 1.5	Va	26/74	96
IVa	$NaNH_2$	THF-EtOH	22°	5.5	Va	50/50	94
IVc	NaOEt	EtOH	r.t.	18	Vb	20/80	79
IVd	NaH	DME	80°	20	Vc	-	55

The above sequence of reactions has now been applied to the direct preparation of permethrin. Thus, the ester IIIa was treated with 3-phenoxybenzyl alcohol under ester-exchange condition to produce 3-phenoxybenzyl 3,3-dimethyl-4-pentenoate (VI), b.p. 155-158°/ 0.3 mm. The BPO catalyzed addition of carbon tetrachloride to the above ester afforded 3-phenoxybenzyl 3,3-dimethyl-4,6,6,6-tetrachlorohexanoate (VII) in 82% yield as a viscous oil, which was purified by column chromatography. The ester VII was then treated with two molar equivalents of sodium t-butoxide in anhydrous THF to give permethrin VIII in 75% yield. The nmr spectrum indicates that it consists of 1:1 mixture of cis and trans isomers.

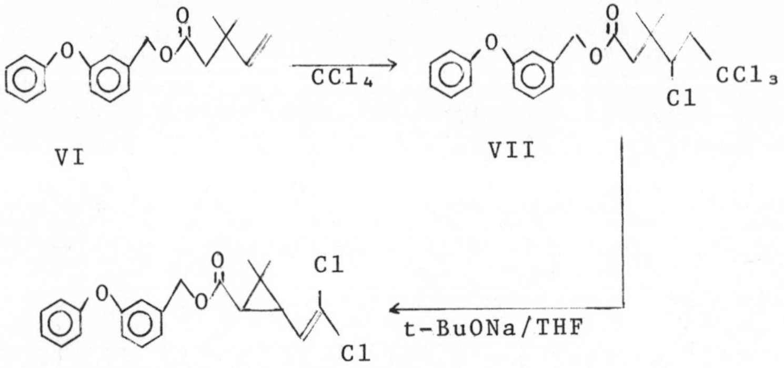

VI $\xrightarrow{CCl_4}$ VII $\xrightarrow{t\text{-}BuONa/THF}$ VIII

At the middle stage of the reaction from IV to V, there was observed the formation of all three possible intermediates, IX, X, and XI.

IX 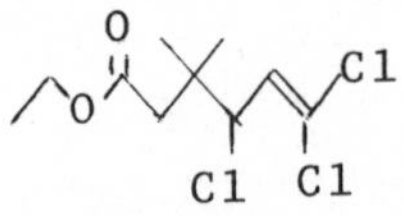 X

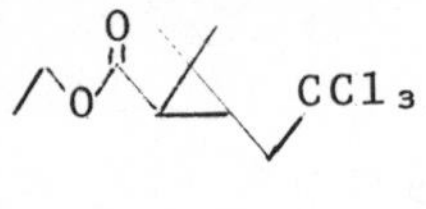

XI 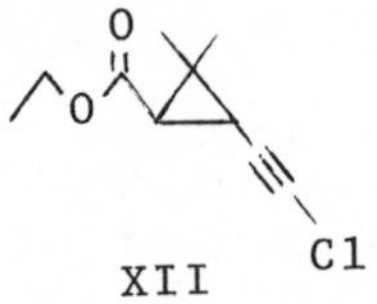 XII

As the reaction proceeds, these intermediates usually disappear, being converted to V. The presence of excess base in the system tends to induce further dehydrohalogenation of V to afford 2,2-dimethyl-3-(2-chloroethynyl)cyclopropanecarboxylate (XII), especially at the final stage of the reaction.

The intermediate IX can selectively be prepared by treatment of the bromotrichloromethane adduct IVb with piperidine in benzene at 80° for 15 hrs. The olefinic linkage in IX was assigned to be trans based on the coupling constant (15 Hz) of the olefinic protons in the nmr spectrum. The reaction of the carbon tetrachloride adduct IVa with pyrrolidine in DMF at room temperature afforded the intermediate X selectively. The selective preparation of the intermediate XI was attained by treating IV with sodium or potassium t-butoxide in hydrocarbon solvent. When IVb was used as starting material, almost 1:1 mixture of cis- and trans-XI was obtained, while IVa was converted predominantly to trans-XI. All these intermediates could be transformed smoothly into V by treatment with base under the same condition being used in the direct conversion of IV to V.

Synthesis of the Homologues of Chrysanthemate

Modification of the dimethylvinyl group in chrysanthemate to 1-propenyl, 1-butenyl, or 1,3-butadienyl substituent also increases significantly the insecticidal activity (15). The acid moieties of these homologues can be prepared by the following sequence of reactions.

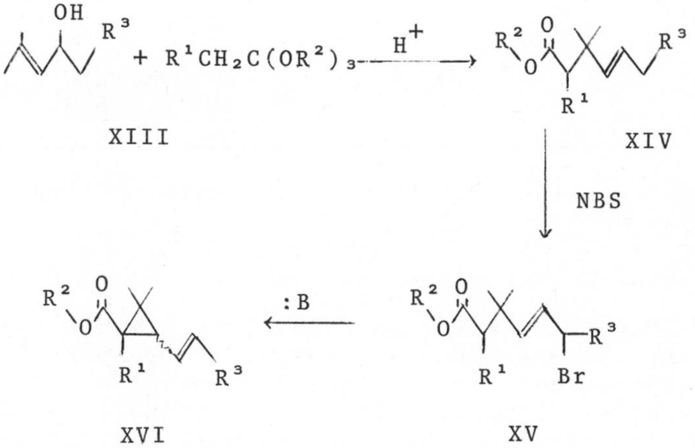

OH R3 + $R^1CH_2C(OR^2)_3$ $\xrightarrow{H^+}$ XIV

XIII XIV

NBS

XVI $\xleftarrow{:B}$ XV

XVI XV

The starting allylic alcohol XIII was prepared either by reduction of mesityl oxide with LAH or by the condensation of 3-methylcrotonaldehyde with appropriate Grignard reagents. Heating of a mixture of the above alcohol XIII and triethyl orthoacetate in the presence of phenol at 140° afforded the γ,δ-unsaturated esters XIV in good yields. The ester XIV was then brominated with N-bromosuccinimide in carbon tetrachloride in the presence of BPO to produce ε-bromo-γ,δ-unsaturated esters XV. Treatment of the resulting ester XV with potassium t-butoxide in THF gave finally the desired cyclopropanecarboxylate XVI. The α-substituted analogues (16) were prepared in a similar manner, as described above, starting with allylic alcohol XIII and triethyl orthopropionate or trimethyl orthobutyrate. The results and the reaction conditions for cyclization step are summarized in Table IV.

Table IV. Homologues of Chrysanthemate

	R^1	R^2	R^3	Product Yield(%) XIV	XV	XVI	Reaction Conditions for Cyclization
a	H	Et	H	64	91	85	t-BuOK/THF 60°C: 4 hr
b	H	Et	Me	85	91*	66**	t-BuOK/THF 0∿5°C: 4 hr
c	H	Et	Et	88	92*	69**	t-BuOK/THF -30°C: 1 hr
d	Me	Et	H	60	86*	77**	t-BuOK/THF -10°C: 1.5 hr
e	Et	Me	H	48	83	22	t-BuOK/THF 0∿5°C: 6 hr

* Crude yields. Crude products were used in next step without purification.

** Isolated yields based on XIV.

The cyclized ester was usually a mixture of cis and trans isomers. For example, the nmr spectrum of the crude product derived from XVa revealed that XVIa was almost a 1:1 mixture of cis and trans isomers. Further treatment of this mixture with t-butoxide in t-butanol at 80°, however, induced the smooth isomerization of the cis isomer to the thermodynamically stable trans XVIa (17). Infrared spectra of the esters XVIb and XVIc exhibited a strong absorption at 960-965 cm^{-1}. Therefore, the geometry of the olefinic bond attached to the cyclopropane ring would be trans, though the presence of cis isomer as a minor component could not be excluded.

Two side-reactions were observed in the cyclization of XV to XVI. The one was the 1,2-elimination of hydrogen bromide leading to diene, which occurred in the reaction with XVb and XVc. The other was the substitution of halogen by t-butoxy anion to give the ester XVII. The latter reaction was observed especially when the α-substituted analogues of XV were exposed to the aforementioned cyclization. The competitive occurrence of these undesired reactions might be the result of insufficient acidity of the α-hydrogen in bromoester XV, especially in XVd and XVe. Both side-reactions, however, could effectively be suppressed by lowering the reaction temperature.

d: R^1=Me, R^2=Et

e: R^1=Et, R^2=Me

XVII

For the syntheses of butadienyl and styryl analogues, γ,δ-unsaturated esters XIVf and XIVg were prepared by the condensation of triethyl orthoacetate with allylic alcohols XIIIf (R^3=vinyl) and XIIIg (R^3=phenyl) in 87 and 75% yields, respectively.

XIVf —NBS→ XVIII —t-BuOK→ XX

XIVg —NBS→ XIX —t-BuOK→ XXI

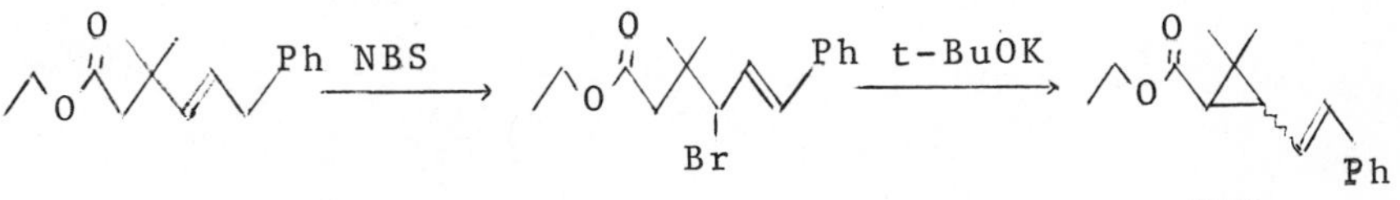

The structure of products obtained by the bromination of these unsaturated esters was not so simple. Inspection of their nmr spectra suggested that the major product from XIVf was the ω-bromoester XVIII (94% yield) and that from XIVg was the γ-bromoester XIX (85% yield). Treatment of these crude bromination products with potassium t-butoxide in THF below 0° produced the expected cyclopropanecarboxylates XX, b.p. 62-65°/0.1 mm., and XXI, b.p. 112-118°/0.1 mm., in 59 and 58% yields based on XIV, respectively.

Acknowledgements. The authors are grateful to Mr. T. Takashima, Mr. T. Koizumi, Mr. K. Sugimoto, and Miss Y. Takahatake for their capable assistance.

Literature Cited.

1. Elliott, M., Farnham, A.W., Janes, N.F., Needham, P.H., Pulman, D.A., and Stevenson, J.H., Nature (1973) 246, 169.
2. Elliott, M., Farnham, A.W., Janes, N.F., Needham, P.H., and Pulman, D.A., Nature (1974) 248, 710.
3. Farkas, J., Kourim, P., and Sorm, F., Coll. Czech. Chem. Commun. (1959) 24, 2230.
4. Crombie, L., Doherty, C.F., and Pattenden, G., J. Chem. Soc. (C) (1970), 1076.
5. Elliott, M., Janes, N.F., and Pulman, D.A., J. Chem. Soc. Perkin I (1974), 2470.
6. Johnson, W.S., Werthman, L., Bartlett, W.B., Brocksom, T.J., Li, T., Faulkner, D.J., and Petersen, M.R., J. Amer. Chem. Soc. (1970) 92, 741.
7. Babler, J.H. and Tortorello, A.J., J. Org. Chem. (1976) 41, 885.
8. Harrison, R.G. and Lythgoe, B., Chem. Commun. (1970), 1513.
9. Bolton, I.J., Harrison, R.G., and Lythgoe, B., J. Chem. Soc. (C) (1971), 2950.
10. Kharasch, M.S., Jensen, E.V., and Urry, W.H., J. Amer. Chem. Soc. (1947) 69, 1100.
11. Kharasch, M.S., Jensen, E.V., and Urry, W.H., J. Amer. Chem. Soc. (1946) 68, 154.
12. Asscher, M. and Vofsi, D., J. Chem. Soc. (1961), 2261.
13. Suzuki, T. and Tsuji, J., Tetrahedron Lett. (1968), 913; J. Org. Chem. (1970) 35, 2982.
14. Matsumoto, H., Nakano, T., and Nagai, Y., Tetrahedron Lett. (1973), 5147.
15. Elliott, M., Farnham, A.W., Janes, N.E., Needham, P.H., and Pulman, D.A., Nature (1973) 244, 456.
16. Itaya, N., Okuno, Y., Horiuchi, F., Higo, A., Honda, T., Mizutani, T., Ohno, N., Kitamura, S., and Matsuo, T., Japan kokai 74-75725, Chem. Abst. (1975) 82, 81723f.
17. Julia, M., Julia, S., Bemont, B., and Tchernoff, G., C. R. Acad. Sci. (1959) 248, 242.

13

Photochemical Reactions of Pyrethroid Insecticides

ROY L. HOLMSTEAD, JOHN E. CASIDA, and LUIS O. RUZO

Pesticide Chemistry and Toxicology Laboratory, Department of Entomological Sciences, University of California, Berkeley, Calif. 94720

The natural pyrethrins and many highly insecticidal synthetic chrysanthemates are not suitable for control of agricultural insect pests because of insufficient stability in light and air (1). Considerable progress has been made in improving the photostability of pyrethroids by suitable formulation (e.g., microencapsulation and inclusion complexes) and by adding antioxidants or UV screens. However, the most effective stabilization is achieved by replacing the photolabile groups by others that give enhanced stability to the overall molecule and equal or increased insecticidal activity (2,3). Knowledge of the photochemical reactions of the earlier pyrethroids contributed to the development of this new generation of photostabilized pyrethroids. It is now necessary to define the photochemistry of these newer compounds and the significance of their photoproducts as residues and environmental contaminants.

This review considers the types of photolytic reactions of pyrethroids with emphasis on permethrin (3), NRDC 161 (4) and S 5602 (5).

Isomerization of the Cyclopropane Ring and of Alkenyl Substituents

Epimerization of [1R]-cyclopropanecarboxylate insecticides greatly reduces or destroys their insecticidal activity. As a corollary, epimerization of suitable [1S]-compounds yields the insecticidal conformation. The *trans*- and *cis*-isomers also differ in potency and persistence. Photoisomerization of the cyclopropane ring therefore has important consequences.

Irradiation ($\lambda > 200$ nm) of [1R]- or [1RS]-*trans*- or -*cis*-chrysanthemic acid or its simple alkyl esters in hexane with isobutyrophenone or related sensitizers yields equilibrium mixtures of the corresponding [1R,*trans*]-, [1S,*trans*]-, [1R,*cis*]- and [1S,*cis*]-compounds in the approximate ratio 32:32:18:18 (6-9) (Figure 1). The postulated mechanism involves cleavage

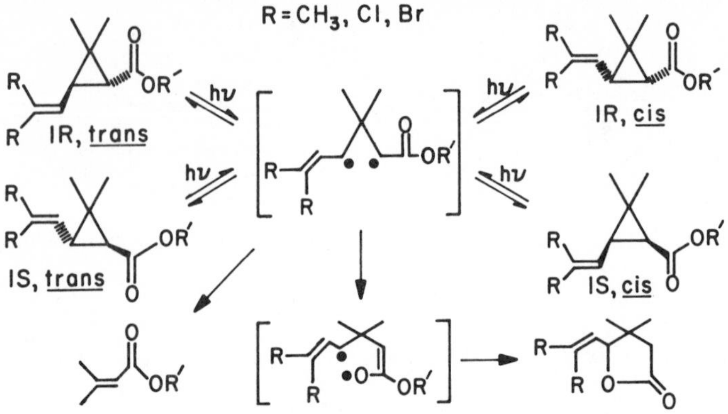

Figure 1

of the C-1 to C-3 bond of the cyclopropane ring to form a diradical which may reform the C-1 to C-3 bond to yield any of the 4 possible isomers. Alternatively, the diradical may fragment by C-2 to C-3 bond cleavage to give the senecioate or it may rearrange to a lactone (7,9).

Studies on 4 insecticidal [1R,trans]-chrysanthemates (pyrethrin I, allethrin, dimethrin and tetramethrin) did not reveal isomerized products even after 24 hr sunlamp irradiation on glass (10). Thus, neither cis-chrysanthemic acid nor meso-cis-caronic acid was recovered on hydrolysis of the photodecomposed esters. The absence of isomerization at the cyclopropane ring was confirmed with trans-resmethrin irradiated on silica gel with sunlight or a sunlamp (11). These chrysanthemates are relatively unstable in light and air so that in short-term studies other reactions may take preference over isomerization of the cyclopropane ring.

Pyrethroids photostabilized in the acid moiety by replacing the isobutenyl group by a dihalovinyl substituent undergo cyclopropane isomerization to a significant extent in both solution and solid phase reactions (12-14). Irradiation ($\lambda > 290$ nm) of either trans- or cis-permethrin leads to an equilibrium mixture of trans- and cis-esters, the isomerization occurring more rapidly in hexane or as thin films on glass than in methanol or as deposits on silica gel or soil surfaces (12). Photoisomerization of permethrin also occurs in thin films exposed to sunlight (12). In dilute aqueous solutions, photoisomerization ($\lambda > 290$ nm) of either trans- or cis-permethrin occurs rapidly with the resulting isomer mixture undergoing further photodecomposition (see below). On irradiation ($\lambda > 290$ nm) of NRDC 161 solutions, a small degree of isomerization occurs in methanol (13) and considerably more in 2-propanol; on exposure of thin films on glass to sunlight, 30% trans-isomer is obtained after 6 days irradiation (14). The senecioate derivatives are also detected on exposure of permethrin and NRDC

161 as thin films to sunlight and of permethrin in water to sunlight (12, 14).

The *trans*/*cis*-isomerization for chrysanthemic acid and its esters occurs *via* the triplet excited state since the isomerization rate is markedly increased with isobutyrophenone and other sensitizers (6-8). This is also the case with the acid moiety of NRDC 161 where isomerization is the only reaction observed on irradiation ($\lambda > 290$ nm) in hexane in the presence of triplet sensitizers but photodebromination is the major reaction without sensitizer (13).

Alkenyl substituents in the rethronyl moiety also undergo photoisomerization, *i.e.*, the allyl group of allethrin is converted to a cyclopropyl substituent and the Z(*cis*)-pent-2-enyl group of jasmolin I gives the E(*trans*)-isomer (15) (Figure 2).

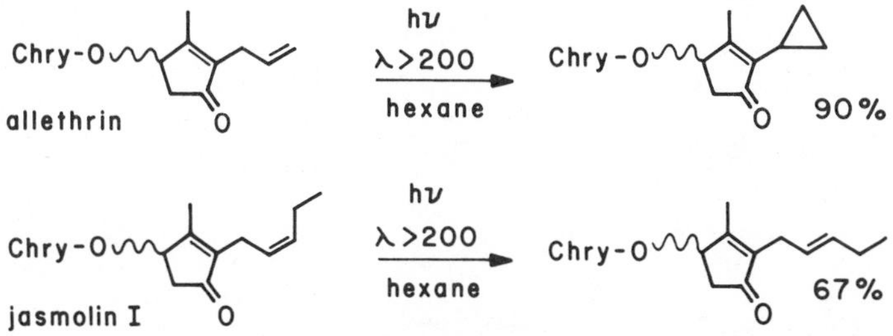

Figure 2

Oxidation of Functional Groups in the Acid and Alcohol Moieties

Photooxidation of the isobutenyl substituent in chrysanthemates and of various functional groups in the alcohol moieties of the earlier pyrethroids greatly limits their residual persistence. Photodecomposition of pyrethrin I, allethrin, dimethrin and tetramethrin as thin films on glass yields 11-15 products in each case (10). Saponification of the mixture of ester products from each pyrethroid liberates 12-16 acids, of which the identified compounds originate from oxidation at the *trans*-methyl group or double bond of the isobutenyl substituent; possible pathways to account for these compounds are shown in Figure 3. Although products derived from the alcohol moiety were not examined, the high lability of pyrethrin I and allethrin relative to the other compounds indicates that the pentadienyl and allyl groups are very susceptible to photooxidation (10).

Resmethrin and other pyrethroids with the 5-benzyl-3-furylmethyl group undergo rapid oxidation when exposed to sunlight or sunlamp irradiation in aqueous medium or as thin films on glass or as deposits on silica gel (11) (Figure 4). One major photodecomposition route involves epoxidation at the isobutenyl substituent to give the R- and S-epoxides. Formation

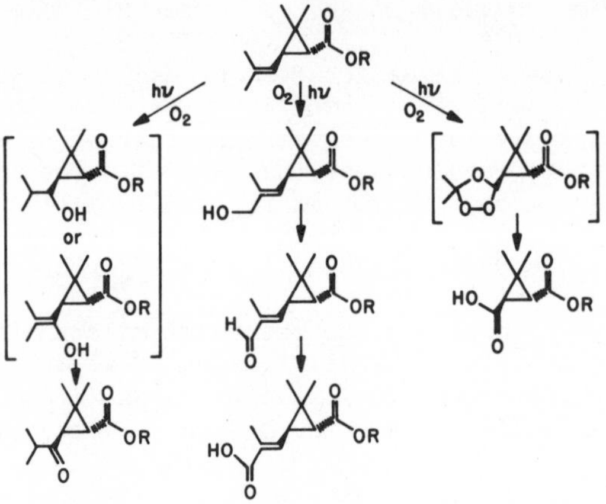

Figure 3

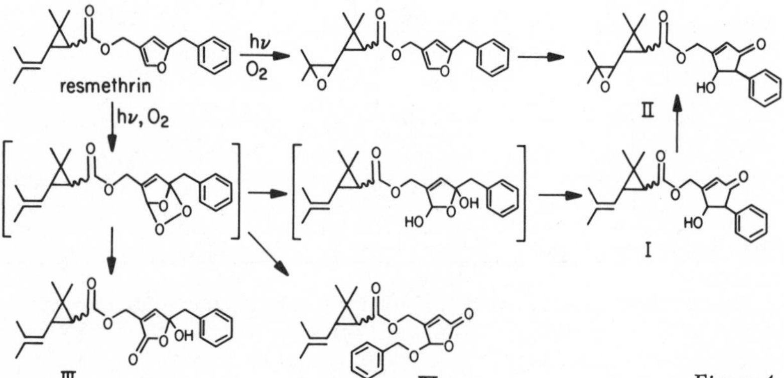

Figure 4

of the other major photoproducts is initiated by oxidation of the furan ring to a cyclic ozonide-type peroxide which decomposes by the following pathways: reduction to a diol followed by rearrangement to the cyclopentenolone (I) which is also detected as its epoxide derivative (II); migration of a proton or a hydrogen radical from the position symmetrical to the benzyl group to give the hydroxy lactone (III); migration of the benzyl cation or radical to give the benzyloxy lactone (IV) (11).

No photooxidized derivatives retaining the ester group are identified to date in the studies on permethrin (12), NRDC 161 (14) and S 5602 (12). The halogen-substituted double bond of the acid moiety and the 3-phenoxybenzyl group appear to be quite resistant to photooxidation.

Reductive Dehalogenation of Dihalovinyl Substituents

The dihalovinyl replacement for the labile isobutenyl function stabilizes the acid group to photooxidation but it introduces the possibility of reductive photodehalogenation to a vinyl halide. Less extensive dehalogenation is expected with permethrin than with NRDC 161 based on their relative carbon-halogen bond strengths.

Permethrin irradiated ($\lambda > 290$ nm) in water gives the monodechlorinated derivative of the parent ester and of the acid moiety but always in minor amounts (12) (Figure 5); somewhat larger amounts of these products are obtained at shorter wavelengths (e.g., $\lambda > 220$ nm). Although the stereochemistry of the

Figure 5

monodechlorinated permethrin is not assigned, it is likely to be the trans-dehalogenated product on analogy with the findings on NRDC 161. Thus, debromination of NRDC 161 in hexane proceeds with a steric preference so that the trans-debrominated product (Figure 5) accounts for ~ 80% of the total monodebrominated derivatives; on prolonged irradiation, a small amount of di-debromo-NRDC 161 is formed (13). This steric preference is not evident starting with the acid moiety of NRDC 161, i.e., the trans/cis debromination ratio approximates unity (13). The presence of a hydrogen donor is necessary for photodebromination of NRDC 161 since debrominated material is not formed on irradiation of either benzene solutions or thin films (14). In the studies carried out to date (12-14), no secondary oxidation products from either monodechloro-permethrin or monodebromo-NRDC 161 have been observed.

Reductive dechlorination of the chlorophenyl group of S 5602 is evident in hexane with light sources containing short wavelength light (12).

Photoelimination of Carbon Dioxide

The importance of photodecarboxylation in ester photolysis generally depends on the structure of the acid and irradiation conditions. Photoelimination of carbon dioxide is a prominent reaction in certain pyrethroids and related compounds which contain an α-cyano or other free radical stabilizing group α to the ester linkage (16). This type of reaction is negligible in unsubstituted esters such as permethrin (12) but it is the major pathway for S 5602 on irradiation ($\lambda > 290$ nm) in methanol (Figure 6) and in hexane, acetonitrile-water and other solvents (12). The decarboxylated derivative is the major photoproduct on

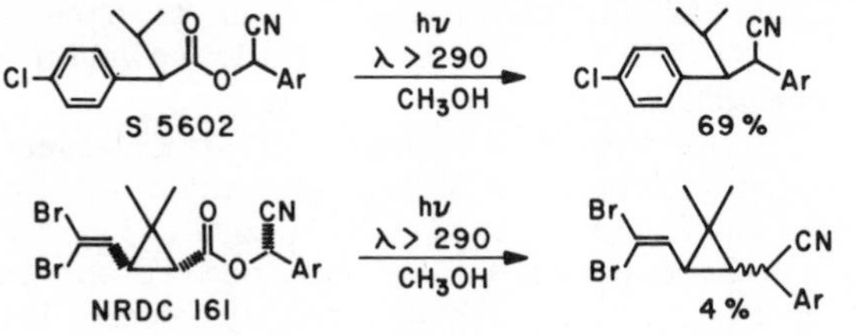

Figure 6

exposure of S 5602 on silica gel to sunlight for 28 days (12). In the case of NRDC 161, the decarboxylation product is formed in methanol (Figure 6) and hexane but in much smaller amount (13) than with S 5602. This difference is probably due to a combination of the stability of the benzylic radical formed from the acid portion of S 5602 and the variety of other reactions available to NRDC 161.

Ester Bond Cleavage

Photolysis of the ester bond is a significant reaction for trans- and cis-resmethrin (11) but apparently not for pyrethrin I, allethrin, dimethrin and tetramethrin (10). It is a major reaction with the newer pyrethroids which contain halogen atoms in the acid moiety and a cyano group at the benzyl carbon. Thus, irradiation of trans-permethrin in water or hexane yields as major products the trans- and cis-dichlorovinyl acids and 3-phenoxybenzyl alcohol whereas photolysis in methanol under similar conditions gives the methyl esters of these acids and the methyl ether of 3-phenoxybenzyl alcohol (12) (Figure 7). cis-

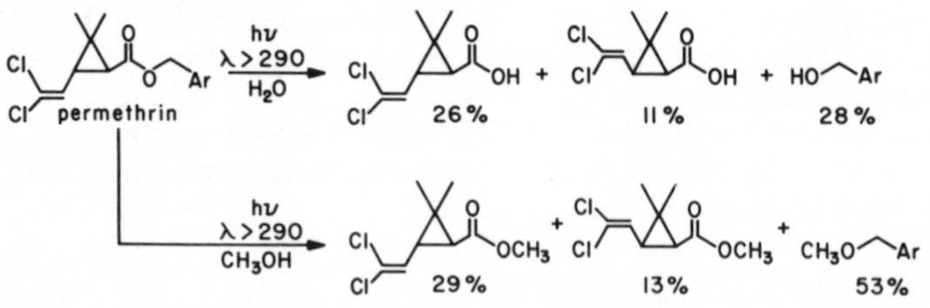

Figure 7

Permethrin undergoes analagous reactions (12). Similarly, NRDC 161 gives the trans- and cis-dibromovinyl acids in hexane and acetonitrile-water and their methyl esters in methanol (13). As the viscosity of the solvent increases (methanol, ethanol and 2-propanol), there is a relative decrease in the extent of ester cleavage with NRDC 161 and in the solid phase, in sunlight, this ceases to be the major reaction pathway (14).

Dimerization of Free Radicals

Irradiation (λ > 290 nm) of NRDC 161 in hexane (13) and of S 5602 in several solvents (12) leads to dimers of free radicals generated during the photolysis process (Figure 8). Dimers

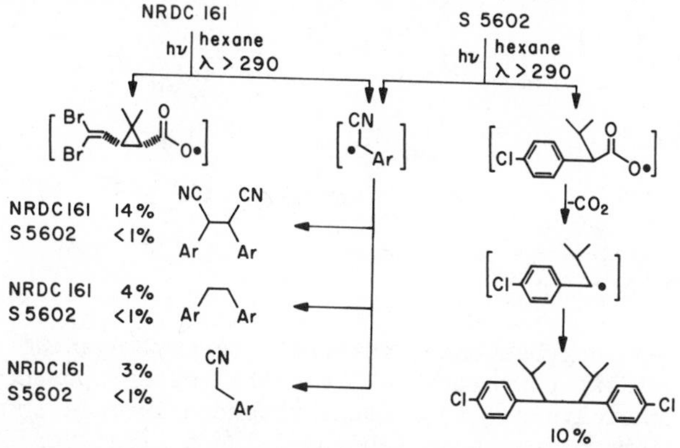

Figure 8

containing the cyano group are presumably formed *via* coupling of the free radicals generated from homolytic cleavage of the oxygen-carbon bond of the alcohol portion of the pyrethroid. The main dimeric product from S 5602 is formed from recombination of the α-isopropyl-*p*-chlorobenzyl radical resulting from the decarboxylation reaction discussed above. This dimer is observed on glc as 2 peaks which give identical mass spectra and probably correspond to the *d,l*-mixture and the *meso* form. Direct photolysis of α-isopropyl-*p*-chlorobenzyl chloride also generates the same dimer.

Further Photodecomposition of Ester Cleavage Products

Pyrethroids generally yield a large number and great variety of photoproducts most of which originate from further reactions of the primary cleavage products. For example, the alcohol moiety liberated on photolysis of resmethrin degrades further to benzyl alcohol, benzaldehyde, benzoic acid and phenylacetic acid, the latter contributing to the unpleasant odor of photodecomposed resmethrin (11). Permethrin photolysis (λ > 290 nm) in water yields 3-hydroxybenzyl alcohol and 3-hydroxybenzaldehyde (12). Several products obtained in varying amounts on photolysis of the α-cyano pyrethroids, NRDC 161 (13) and S 5602 (12), are shown in Figure 9. Small amounts of the free cyanohydrin are observed on NRDC 161 photolysis in hexane, methanol and water and large amounts of 3-phenoxybenzaldehyde are formed with both pyrethroids in hexane and methanol. The benzoyl cyanide is a major product in hexane whereas in methanol it reacts further to

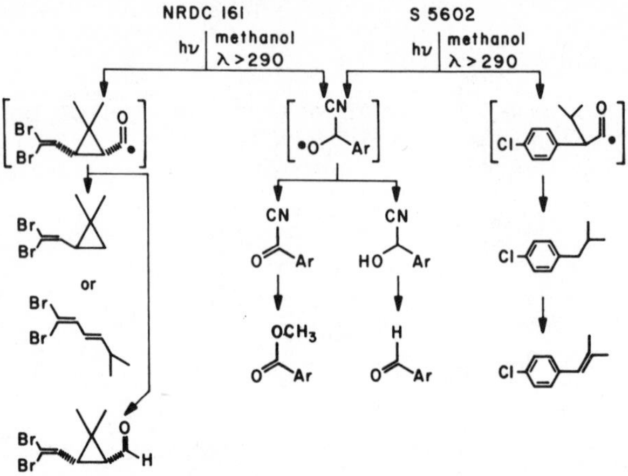

Figure 9

give methyl 3-phenoxybenzoate; authentic benzoyl cyanide in methanol is readily converted to the methyl ester upon heating at 50° or on photolysis (13). Other products include the aldehyde of the dibromovinyl acid (which yields a great variety of additional products on further photolysis) and the dibromovinylcyclopropane derivative or its ring-opened isomer. Analogous reactions occur in the acid moiety on photolysis of S 5602.

Discussion

The first steps have been taken in understanding pyrethroid photochemistry, a field that will undoubtedly undergo tremendous growth within the next few years. This knowledge is useful in further stabilizing the pyrethroids to photodecomposition but it also signals certain peripheral problems. The early pyrethroids were too unstable in light and air for extensive use in agriculture whereas currently available pyrethroids are sufficiently stable so that weekly or biweekly applications provide excellent pest insect control. Further stabilization may increase the risk of unfavorable environmental persistence. The large number and great variety of photoproducts provide a challenge to analytical chemists and toxicologists responsible for devising methods of residue analysis and experiments to evaluate the use safety of these highly effective insecticides.

Abstract

Natural and synthetic pyrethroids undergo one or more of the following types of reactions upon photolysis in organic solvents (hexane, methanol), in water or as thin films: isomerization of the cyclopropane ring and of alkenyl substituents; oxidation of functional groups in the acid and alcohol moieties; reductive dehalogenation of dihalovinyl substituents; photoelimination of carbon dioxide, particularly with α-cyanobenzyl compounds; ester bond cleavage yielding the free acid and alcohol moieties; dimerization of free radicals generated during the photolysis process; further photodecomposition of ester cleavage products. The relative importance of these reactions is dependent upon the structure of the pyrethroid and the photolysis conditions.

Acknowledgements

The authors thank Donald Fullmer and Tadaaki Unai for valuable suggestions and assistance. This study was supported in part by grants from: National Institutes of Health (2 P01 ES00049); Agricultural Chemical Div., FMC Corp., Middleport, N.Y.; Agricultural Chemicals Div., ICI United States Inc., Goldsboro, N.C.; Sumitomo Chemical Co., Osaka, Japan; Roussel-Uclaf-Procida, Paris, France; Mitchell Cotts & Co. Ltd., London, England; Wellcome Foundation Ltd., London, England; National Research Development Corp., London, England.

Literature Cited

1. Elliott, M., in "Pyrethrum the Natural Insecticide" (Casida, J. E., Ed.) (1973) Academic Press, New York, N.Y., p. 55 ff.
2. Elliott, M., Farnham, A. W., Janes, N. F., Needham, P. H., Pulman, D. A., Stevenson, J. H., Proc. Seventh Br. Insec. Fung. Conf. (Brighton) (1973) 721.
3. Elliott, M., Farnham, A. W., Janes, N. F., Needham, P. H., Pulman, D. A., Stevenson, J. H., Nature (1973) 246, 169.
4. Elliott, M., Farnham, A. W., Janes, N. F., Needham, P. H., Pulman, D. A., Nature (1974) 248, 710.
5. Matsuo, T., Itaya, N., Mizutani, T., Ohno, N., Fujimoto, K., Okuno, Y., Yoshioka, H., Agr. Biol. Chem. (1976) 40, 247.
6. Sasaki, T., Eguchi, S., Ohno, M., J. Org. Chem. (1968) 33, 676.
7. Sasaki, T., Eguchi, S., Ohno, M., J. Org. Chem. (1970) 35, 790.
8. Ueda, K., Matsui, M., Tetrahedron (1971) 27, 2771.
9. Bullivant, M. J., Pattenden, G., Pyrethrum Post (1971) 11(2), 72.
10. Chen, Y-L., Casida, J. E., J. Agr. Food Chem. (1969) 17, 208.
11. Ueda, K., Gaughan, L. C., Casida, J. E., J. Agr. Food Chem. (1974) 22, 212.
12. Holmstead, R. L., unpublished results.

13. Ruzo, L. O., Holmstead, R. L., Casida, J. E., Tett. Lett. (1976) 35, 3045.
14. Ruzo, L. O., unpublished results.
15. Bullivant, M. J., Pattenden, G., J. Chem. Soc. (1976) 249.
16. Holmstead, R. L., Fullmer, D. G., J. Agr. Food Chem. (1976) accepted for publication.

14

Permethrin Degradation in Soil and Microbial Cultures

DONALD D. KAUFMAN and S. CLARK HAYNES

Agricultural Environment Quality Institute, Agricultural Research Service, U.S. Department of Agriculture, Beltsville, Md. 20705

EDWARD G. JORDAN and ANTHONY J. KAYSER

Department of Botany, University of Maryland, College Park, Md. 20742

Pyrethroids are one of the oldest classes of organic insecticides known. Although natural and synthetic pyrethroids are excellent insecticides, their instability in light and air has limited their use in protecting agricultural crops. Recent work (1) has demonstrated, however, that the most labile groups in pyrethroids can be replaced by others which provide greater stability and equal or increased insecticidal activity.

Knowledge of the pathways by which natural and synthetic pyrethroids are metabolized in mammals (2-11), or photochemically degraded (1, 12-14), has developed rapidly in the last several years. A literature survey indicated that despite their long history of use, essentially nothing is known about the degradation or persistence of pyrethroids in soil. This paper describes the results of a cursory investigation of the degradation and persistence of permethrin [m-phenoxybenzyl cis, trans-(±)-3-(2,2-dichlorovinyl)-2,2-dimethylcyclopropanecarboxylate] (FMC 33297, NRDC 143) in soil (15). A more detailed report will be published elseshere.

Degradation in Aerobic Soil

Aerobic soil metabolism studies were performed with soils placed in a simple flow-through system which permits simultaneous measurement of loss by volatilization and metabolic CO_2 evolution from soil (16). Chemical and physical characteristics of the soils used are listed in Table 1. ^{14}C-Carbonyl (acid) and ^{14}C-methylene (alcohol) permethrin (Fig. 1) were used in these investigations. Material applications were made in 0.1 ml benzene to a final concentration of 0.2 lb/A of the cis/trans mixture after which each sample was thoroughly mixed, watered to 75% moisture content at 1/3 bar moisture, and incubated at 25°C. Sodium azide was used as a microbial inhibitor in soils to assess the contribution of soil microbial activity to permethrin degradation. At the conclusion of the incubation period, the soils were extracted and processed as shown in Fig. 2.

Figure 1. ^{14}C-labeling pattern of ^{14}C-permethrin indicating carbonyl and methylene labeling positions

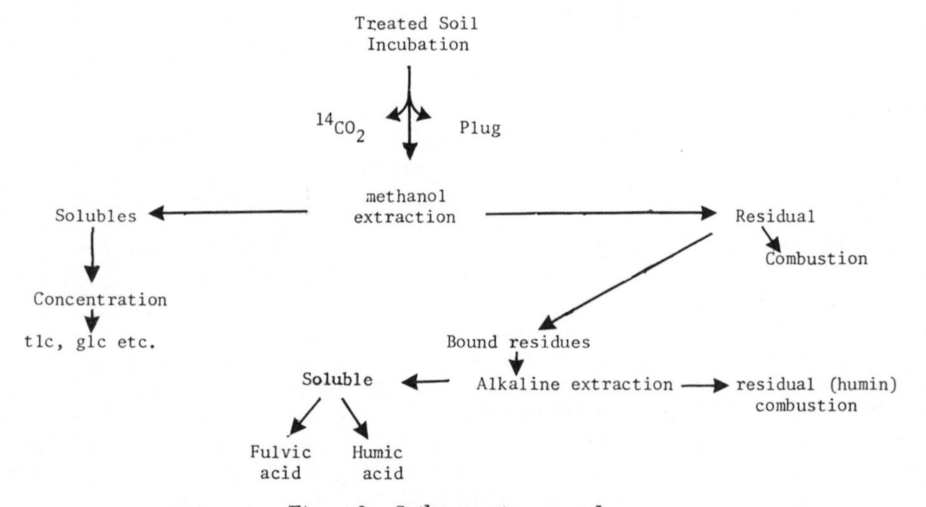

Figure 2. Soil extraction procedure

Table I. Chemical and physical characteristics of soils.

Soil type	CEC (meq/100gm)	% Sand	% Silt	% Clay	% O.M.	pH	% Mois. content 1/3 bar
Memphis silt loam	16.3	20.8	54.0	25.2	0.7	5.8	37.6
Dubbs loam	8.5	48.8	44.0	7.2	1.0	5.9	23.7
Sharkey Clay	33.6	20.8	32.0	47.2	6.1	5.9	45.5
Hagerstown sil. clay loam	8.8	17.0	50.6	32.4	2.3	7.5	32.6
San Joaquin sandy loam		48.0	42.0	9.7	1.2	7.2	22.4

Degradation with subsequent evolution of $^{14}CO_2$ from both ^{14}C-carbonyl- and ^{14}C-methylene-permethrin occurred rapidly in the Hagerstown silty clay loam (Fig. 3). In an initial experiment 62% of the ^{14}C from the methylene, and 52% from the carbonyl-labeled permethrin had been evolved as $^{14}CO_2$ (Fig. 3A) after 27 days incubation. These final differences were reversed in another experiment (Fig. 3B), where 64.5% of the ^{14}C-carbonyl- and 58.7% of the ^{14}C-methylene permethrin had been evolved as $^{14}CO_2$ after 34 days incubation. In both experiments $^{14}CO_2$ evolution was initially more rapid from the ^{14}C-methylene labeled material than from the ^{14}C-carbonyl permethrin. It is doubtful at this time, however, that these differences can actually be considered significant.

The influence of high concentrations of the microbial inhibitor sodium azide on $^{14}CO_2$ evolution from the permethrin treated soils was also examined. Less than 0.3% of the ^{14}C-permethrin was evolved from the azide treated soils. These results indicate that soil microbial activity is involved in the degradation and ultimate evolution of $^{14}CO_2$ from both forms of ^{14}C-permethrin.

Less than 1% of the ^{14}C activity initially present in the soil was recovered as volatile products trapped by the polyurethane plugs. These results indicate that negligible losses of either permethrin itself or degradation products containing either label would occur by volatilization when permethrin is incorporated into soil.

Data obtained from the soil extraction procedure are presented in Table 2. These data indicate that a nearly complete ^{14}C-balance was obtained with both the ^{14}C-methylene- and ^{14}C-carbonyl-permethrin treated soils. Only 14.5-18.8% of the residual ^{14}C activity was removed by methanol extraction from the soil with no microbial inhibitor. Nearly 70% of the ^{14}C residues were extractable from the azide treated soils. Approximately 23-33% of the ^{14}C activity remained associated with the soil residual material (fulvic and humic acids, and humin).

Slight differences were observed in the distribution of the ^{14}C-activity within soil organic matter fractions. The bulk of the ^{14}C activity from both labeled permethrin forms was present in fulvic acid and humin. A somewhat greater portion of ^{14}C-carbonyl permethrin residue was present in fulvic acid than in the humin, whereas a distinctly greater portion of the ^{14}C-methylene permethrin residue was in the humin fraction. In azide treated soils the bulk of both ^{14}C-labels was present in the fulvic acid fraction. These results indicate the importance of soil microbial activity in the degradation of pesticides and the association of the various products with the various soil organic matter fractions.

Significant differences in rates of degradation of ^{14}C-carbonyl permethrin were observed in experiments with different soil types (Fig. 4). Rapid degradation as evidenced by $^{14}CO_2$ evolution was observed in Hagerstown silty clay loam and Dubbs fine sandy loam, whereas intermediate rates of degradation were observed in Sharkey clay and Memphis silt loam. A very slow degradation rate

Table II. ^{14}C-Balance in ^{14}C-permethrin treated Hagerstown silty clay loam

^{14}C-Label position	% ^{14}C recovered as				
	Volatiles		Extract-		
	$^{14}CO_2$	Plug	able	Residual	Total
Carbonyl	64.5	0.2	18.8	27.3	110.8
Carbonyl + NaN_3	0.3	0.4	71.5	22.9	95.1
Methylene	58.7	0.1	14.5	25.4	98.7
Methylene + NaN_3	0.1	0.1	67.7	32.6	100.5

Table III. ^{14}C-Distribution in soil organic matter (Bound residue)

^{14}C-Label position	Soil organic matter fraction		
	Fulvic	Humic	Humin
Carbonyl	48.6	9.7	41.6
Carbonyl + NaN_3	70.5	3.4	26.1
Methylene	32.3	15.7	52.1
Methylene + NaN_3	54.3	13.8	31.9

Table IV. ^{14}C-Balance in ^{14}C-carbonyl-permethrin treated soils

Soil type	% ^{14}C recovered as				
	Volatiles				
	$^{14}CO_2$	Plug	Extractable	Residual	Total
San Joaquin sandy loam	2.2	0.6	86.7	10.6	100.1
Dubbs fine sandy loam	46.0	0.7	17.1	38.7	102.5
Memphis silt loam	31.5	2.4	18.6	45.0	97.5
Hagerstown silty clay loam	51.0	0.4	22.5	26.0	99.9
Sharkey clay	31.1	0.3	40.7	28.5	100.6

was observed in the San Joaquin sandy loam: only 2.2% of the ^{14}C-permethrin was evolved as $^{14}CO_2$ in 28 days. Whether or not this particularly slow rate of permethrin degradation in this soil is associated with some unique microbiological differences or other soil chemical or physical characteristics is not presently known. Some possible insights, however, were obtained by analysis of the soil extracts.

Exceptionally good ^{14}C-balances were obtained in each of the five soils (Table 4). Nearly 87% of the ^{14}C-permethrin was extractable from the San Joaquin soil. Intermediate amounts were extracted from the Hagerstown and Sharkey soils, whereas only lower amounts were obtained from the Dubbs and Memphis soils. Again, as in the previous experiments, the bulk of the residual ^{14}C-activity appeared to be associated with either the fulvic acid and humin soil organic matter fractions (Table 5).

Degradation in Anaerobic Soils

Anaerobic degradation of ^{14}C-permethrin was examined in flooded Hagerstown silty clay loam. The treated soils were contained in soil biometer flasks (17) containing a nitrogen atmosphere. In contrast to aerobic soils, less than 1% of the ^{14}C introduced into the system was trapped as $^{14}CO_2$ from anaerobically incubated soils. Whether or not more was present in the aqueous phase of the flooded system was not determined. Based on the high total recoveries of ^{14}C from the treated soils, however, it would seem doubtful that much of the ^{14}C-label had been converted to $^{14}CO_2$.

Several additional contrasts are also of interest. The total extractable ^{14}C-products were much greater from anaerobically incubated soils (Table 6) than from aerobically incubated soils (Tables 2 & 4). There also appeared to be an increased residual activity in the aqueous phase from the 30 days incubation to the 60 days incubation (Table 6). This would suggest a trend toward more polar product formation. Also, as in aerobically incubated soil, the major portion of the residual ^{14}C in anaerobic soil is associated with the fulvic acid and humin fractions (Table 7). The distribution within these two fractions, however, appears to vary with isotope and incubation period. Hexane was used as the initial extractant in order to trap and remove any possible ^{14}C-methane formed during ^{14}C-permethrin degradation. Although slightly more ^{14}C activity was observed in the hexane extracts of the ^{14}C-methylene permethrin treated soils, confirmation of ^{14}C-methane awaits further analysis.

Soil Product Identification

Thin layer chromatographic analysis of soil extracts from ^{14}C-permethrin treated soil revealed that the parent material is rapidly degraded in soil to a number of products. Tentative identification of three products was established by TLC. These are 3-(2,2-dichlorovinyl)-2,2-dimethylcyclopropanecarboxylic acid,

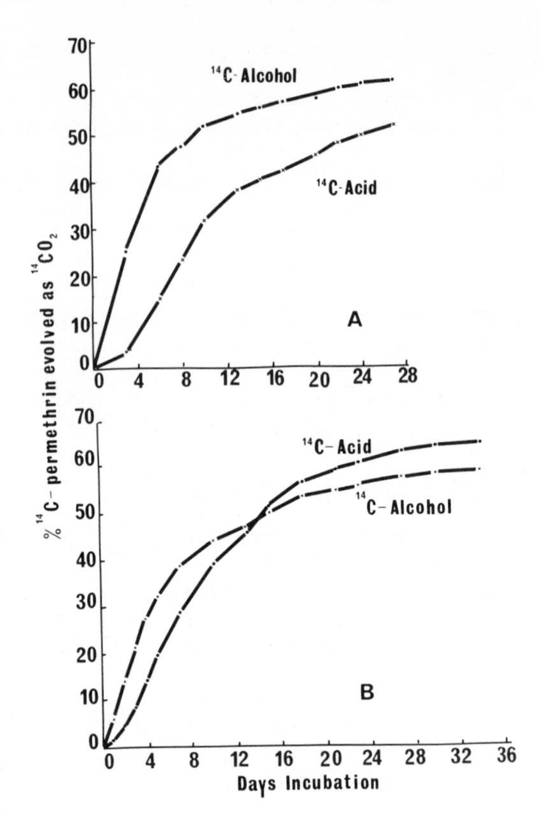

Figure 3. Degradation of ^{14}C-carbonyl (acid) and C^{14}-methylene (alcohol) permethrin in Hagerstown silty clay loam

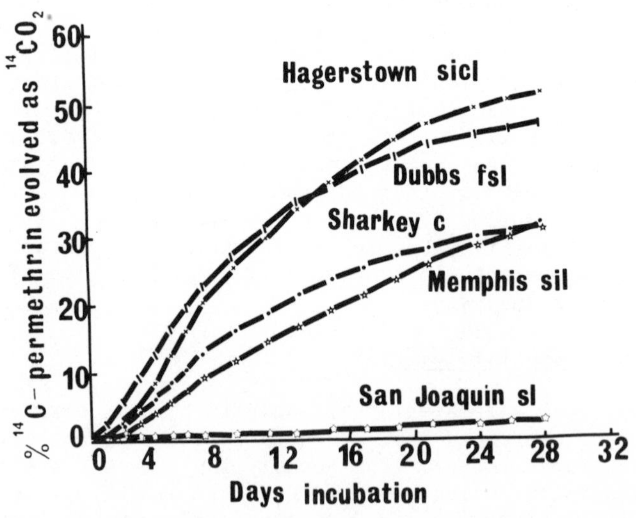

Figure 4. Degradation of ^{14}C-carbonyl permethrin in five soils

Table V. Distribution of ^{14}C from ^{14}C-carbonyl-permethrin in soil organic matter (Bound residue)

Soil type	% ^{14}C present in		
	Fulvic	Humic	Humin
San Joaquin sandy loam	81.2	0.5	18.3
Dubbs fine sandy loam	50.2	5.0	44.9
Memphis silt loam	64.6	1.2	34.2
Hagerstown silty clay loam	38.2	7.7	54.2
Sharkey clay	40.2	2.9	56.9

Table VI. ^{14}C-Distribution from ^{14}C-permethrin in anaerobic Hagerstown silty clay loam

^{14}C-label position	% ^{14}C recovered as					
		Extractable				
	^{14}CO	Hexane	Chloroform Methanol	Aqueous	Residual	Total
30-days incubation						
Carbonyl	0.2	8.2	58.2	n.d.*	27.2	93.8
Methylene	0.1	12.4	81.0	3.7	12.2	109.4
60-days incubation						
Carbonyl	0.3	6.8	64.4	23.2	2.9	97.6
Methylene	0.1	12.6	48.4	11.6	12.6	85.3

n.d. = not determined

3-phenoxybenzyl alcohol, and 3-phenoxybenzoic acid. These products and the parent materials generally appeared as the predominant radioproducts on TLC plates. Numerous other ^{14}C-products also were detected. These products, however, generally occurred in quantities less than 1% of the total products isolated.

For purposes of discussion the C-compounds appearing on 2-dimensional plates were divided into 5 categories (A-E) based on progressively increasing polarity (Fig. 5). Two or three compounds appear to be present in category A, the least polar category. These compounds are less polar than either the cis or trans permethrin which appear in category B. In the experiments described herein, they contain only the carbonyl label, and represent 0-1.5% of the products from either aerobic or anaerobic soils. Cis and trans permethrin are the only compounds present in category B, and generally represented 66-94% of the ^{14}C-materials recovered from aerobic soil, and 43 to 84% of the ^{14}C-materials from anaerobic soil. Two to 9 compounds comprised category C which was intermediate in polarity between the compounds of group D and B. These compounds were present in quantities of 0.5% to 1.5% of the ^{14}C activity extracted. The solvent systems used in this investigation were only capable of partially separating the 3-phenoxybenzyl alcohol and acid, and the dichlorovinyl acid which comprised category D. These products, however, were all distinct from other degradation products and standards examined. They are intermediate in their polarity and chromatograph midway between the less polar cis, trans permethrims and the origin. They contained 2-20% of the ^{14}C activity extracted from aerobic soil, and 2-51% of the activity from anaerobic soils.

Category E contained the most polar compounds which exhibited little movement from the origin. Compounds appearing in this category were frequently unique to the methylene label or both the carbonyl and methylene labels, but seldom to the carbonyl label alone. Considerable additional work is needed to further characterize and/or identify many of the more minor metabolites.

Since all extracts were concentrated to a known volume and a standard amount was used for TLC work, it was possible to make approximate quantitations of the residual permethrin present in the soils at the time of extraction. These data are presented in Table 8 for soil metabolism experiments. Good agreement between labeling patterns was obtained in experiments employing both labeled forms of permethrin. These data indicate that, with the exception of the San Joaquin soil, the 1/2-life of permethrin in soil appears to relatively short i.e., less than 28 days. Additional detailed work, however, will be needed to more accurately determine this figure.

Since the chromatographic systems used provided excellent separation of the cis and trans isomers of permethrin, it was also possible to determine their individual rates of dissipation and the resulting changes in the cis/trans ratio of the residue. Such a calculation, however, assumes no major differences in adsorptive characteristics or in their extractability. The calculated cis/-trans ratios of the extracted permethrin are presented in Table 9.

Table VII. Distribution of ^{14}C in soil organic matter fractions of anaerobically incubated soils.

^{14}C Label position	% ^{14}C in soil organic fraction		
	Fulvic	Humic	Humin
30 days incubation			
Carbonyl	24.7	4.3	71.0
Methylene	33.4	11.2	55.4
60 days incubation			
Carbonyl	68.5	10.6	20.9
Methylene	55.9	19.9	24.3

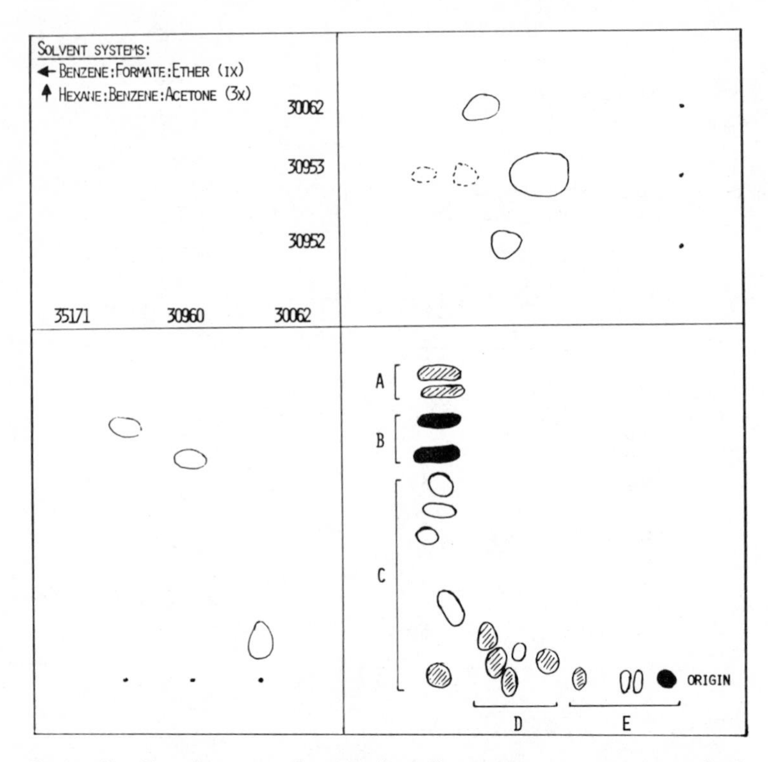

Figure 5. Two-dimensional TLC plate illustrating positions of standards and categories (A–E) *of* ^{14}*C-soil metabolic products generally isolated from* ^{14}*C-permethrin-treated soil. Solid spots:* ^{14}*C-products common to all* ^{14}*C-labels; shaded spots: significant* ^{14}*C-products of specific* ^{14}*C-labels; open spots: minor* ^{14}*C-products. Standards: 30062, dichlorovinyl acid; 30952, 3-phenoxybenzyl acid; 30953, 3-phenoxybenzyl alcohol; 30960,* trans-*permethrin; 35171,* cis-*permethrin.*

Table VIII. % cis,trans-Permethrin remaining after indicate incubation period.

Experiment	Incubation (days)	% ^{14}C-permethrin remaining	
		carbonyl	methylene
Aerobic I	27	7.3	7.6
Aerobic II nonsterile	34	11.4	11.3
sterile	34	65.3	57.7
Anaerobic	30	38.0	25.5
	60	16.0	10.7
Five soils			
Dubbs fsl	28	6.9	
San Joaquin sl	28	58.0	
Memphis sil	28	15.1	
Hagerstown sicl	28	15.5	
Sharkey clay	28	27.7	

Table IX. Cis/trans ratio of residual ^{14}C-permethrin in soil experiments.

Aeration status	cis/trans ratio of ^{14}C-permethrin	
	carbonyl	methylene
Original isotope	46/54	22/78
Aerobic I	64/36	30/70
II nonsterile	60/40	30/70
sterile	56/44	28/72
Anaerobic 30 days	62/38	40/60
60 days	57/43	31/69
Five soils		
Dubbs fsl	64/36	
San Joaquin sl	57/36	
Memphis sil	71/29	
Hagerstown sicl	57/43	
Sharkey clay	67/43	

Since the *trans* figure is the decreasing figure, it can be assumed that this *isomer* is the most rapidly and easily degraded isomer. This observation would be consistent with the work of others (6,7) which indicates that the *trans* isomer is more readily hydrolyzed.

Microbial Degradation

A comparison of results obtained in nonsterile and sodium azide treated soils (Tables 2 & 3) indicate that soil microorganisms play a role in the degradation of permethrin in soil. Two enrichment experiments were performed to attempt observing microbial degradation of permethrin in culture solutions, and to isolate into pure culture soil microorganisms capable of metabolizing permethrin as a source of carbon or energy. In an initial enrichment experiment, each flask was inoculated with 5 g Hagerstown silty clay loam. In a second enrichment experiment, each flask was inoculated with 5 ml of an aqueous soil suspension. In a third experiment, 5 ml of an actively growing cell suspension of *Fusarium oxysporum* Schlecht was used as inoculum. This organism was selected for its known esterase-amidase type activity (18). All culture solutions were monitored for liberation of free chloride ion, evolution of volatile permethrin degradation products, and $^{14}CO_2$ evolution.

Essentially no chloride ion liberation or $^{14}CO_2$ evolution occurred in microbial culture solutions containing either ^{14}C-methylene or ^{14}C-carbonyl permethrin. Although it is conceivable that the permethrin concentrations used in these investigations were toxic to permethrin degrading microorganisms, the culture solution did support luxuriant growth of a wide variety of microorganisms in the presence of these concentrations. At the conclusion of these experiments the soil and/or cell growth were removed from the culture medium by filtration. Both the culture medium and the culture filtrates were extracted. The major portion of the ^{14}C activity was found associated with the soil residue or cell debris (Table 10).

Examination of culture extracts by TLC revealed that some microbial degradation had indeed occurred. However, essentially no changes occurred in the *cis/trans* ratio of the permethrin in culture solutions. This is particularly interesting in view of the results obtained in soil experiments where changes were more pronounced. This would suggest that under the conditions of these microbial experiments, there was very little isomer effect observed, and that the isomers were hydrolyzed at about the same rate. Further investigation will be necessary to further substantiate this phenomenon.

Summary

The results of these investigations indicate that permethrin is rapidly degraded in soil. Degradation of the *trans* isomer occurs more rapidly than with the *cis* isomer. The results

Table X. ^{14}C-Balance in ^{14}C-permethrin microbial metabolism experiments.

Inoculum	Label	% ^{14}C recovered from						Total
		Volatiles		Extractable (media)	Aqueous Residual	Extractable (filtrate)	Residual Combustion	
		CO_2	Plug					
Enrichment								
5 gm soil	Carbonyl	0.4	2.8	6.0	0.4	66.5	0.9	77.0
	Methylene	2 1	0.0	2.4	0.1	67.8	2.2	74.6
5 ml soil extract	Carbonyl	1.1	0.1	-	13.7	51.3	2.9	69.1
	Methylene	1.2	0.1	-	22.7	39.7	5.0	68.7
Pure culture								
F. oxysporum	Carbonyl	0.2	0.2	0.9	1.6	79.6	1.7	84.2
	Methylene	0.2	0.0	1.1	6.2	78.2	0.9	86.6

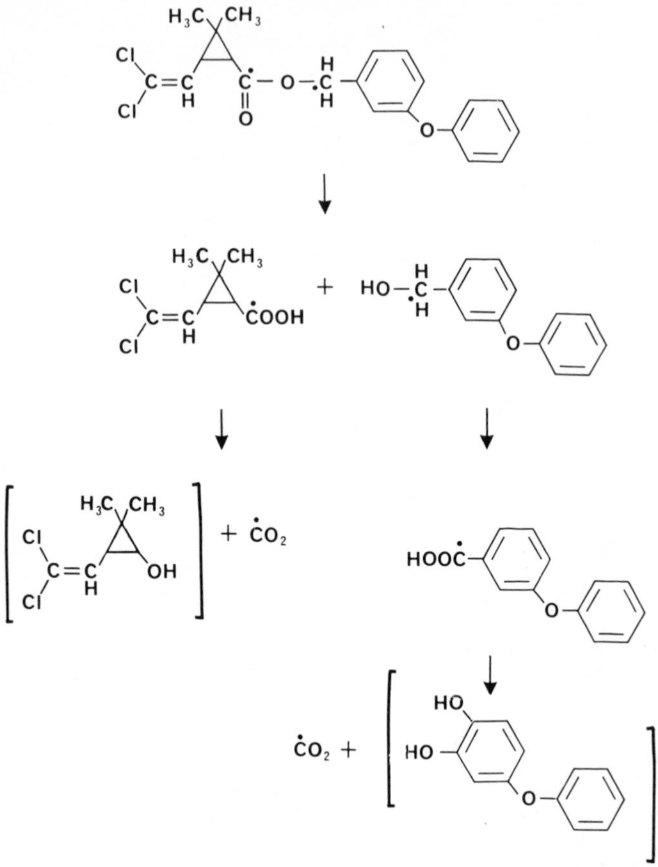

Figure 6. Tentative degradation pathway of permethrin in soil

obtained with enrichment and isolated cultures, and in soil experiments comparing nonsterile and sodium azide treated soils indicate that microbial metabolism is also involved.

Based on results obtained in identifying soil degradation products, it is apparent that the major degradation mechanism of permethrin is hydrolysis to the dichlorovinyl acid and 3-phenoxybenzyl alcohol moieties (Fig. 6). Further metabolism of both products results in the evolution of the ^{14}C-label as $^{14}CO_2$. Based on these findings, the tentative pathway shown in Fig. 6 is proposed. The conversion of 3-phenoxybenzyl alcohol to 3-phenoxybenzoic acid presumably occurs through the formation of the corresponding aldehyde. Additional investigation is needed to further substantiate this pathway. Other possible products would presumably include various hydroxylated parent compounds, as well as hydroxylated cleavage products.

Literature Cited

1. Elliot, M., Farnham, A. W., James, N. W., Needham, P. H., Pulman, D. A., Stevenson, J. H. Nature (London) (1973), 246, 169.
2. Abernathy, C. O., Veda, K., Engel, J. L., Gaughan, L. C., Casida, J. E., Pestic. Biochem. Physiol. (1973), 3, 300.
3. Casida, J. E. in "Pyrethrum the Natural Insecticide," Casida, J. E., Ed., Academic Press, New York, N. Y., (1973) pp 101-120.
4. Casida, J. E., Veda, K., Gaughan, L. C., Jas, L. T., Soderlund, D. M., Arch. Environ. Contam. Toxicol. in press (1976).
5. Elliot, M., Janes, N. F., Kimmel, E. C., Casida, J. E., J. Agric. Food Chem. (1972), 20, 300.
6. Miyamoto, J., Nishida, T., Veda, K., Pestic. Biochem. Physiol. (1971), 1, 293.
7. Miyamoto, J., Suzuki, T., Nakae, C., Pestic. Biochem. Physiol. (1974), 4, 438.
8. Suzuki, T., Miyamoto, J., Pestic. Biochem. Physiol., (1974) 4, 86.
9. Veda, K., Gaughan, L. C., Casida, J. E., J. Agric. Food Chem. (1975a), 2, 106.
10. Veda, K., Gaugham, L. C., Casida, J. E., Pestic. Biochem. Physiol., (1975b), 5, 280.
11. Elliott, M., Janes, N. F., Pulman, D. A., Gaugham, L. C., Vnai, T., and Casida, J. E., J. Agric. Food Chem., (1976), 24, 270.
12. Chen, Y.-L., Casida, J. E., J. Agric. Food Chem., (1969), 17, 208.
13. Elliot, M., Janes, N. F., in "Pyrethrum the Natural Insecticide, "Casida, J. E., Ed., Academic Press, New York, N.Y., (1973), p. 86.
14. Veda, K., Gaughan, L. C., Casida, J. E., J. Agric. Food Chem. (1974), 22, 212.
15. Kaufman, D. D., Jordan, E. G., Abst. 172nd ACS Mtg. Pestic. Chem. Div., (1976), No. 33 (San Francisco, Calif.)
16. Kearney, P. C., Konston, A., J. Agric. Food Chem., (1976), 24, 424.
17. Bartha, R., Pramer, D., Soil Sci. (1965), 100, 68.
18. Blake, J., Kaufman, D. D., Pestic. Biochem. Physiol., (1975), 5, 305-313.

15

Substrate Specificity of Mouse-Liver Microsomal Enzymes in Pyrethroid Metabolism

DAVID M. SODERLUND[1] and JOHN E. CASIDA

Pesticide Chemistry and Toxicology Laboratory, Department of Entomological Sciences, University of California, Berkeley, Calif. 94720

Rapid detoxification contributes to the low acute and chronic toxicity to mammals of the pyrethrins and other chrysanthemates (1,2). Increased insecticidal potency in the newer synthetic pyrethroids has been achieved by replacing some of the biodegradable groupings by substituents that retain overall insecticidal configurations but are more refractory to metabolism (3). Currently important pyrethroids include 9 acid moieties [A-I; shown as 1R,*trans* (A-F) or most insecticidal isomer (G, I)] and 9 alcohol moieties [a-i; the most insecticidal isomer of a-c and h is shown] as follows (Figure 1):

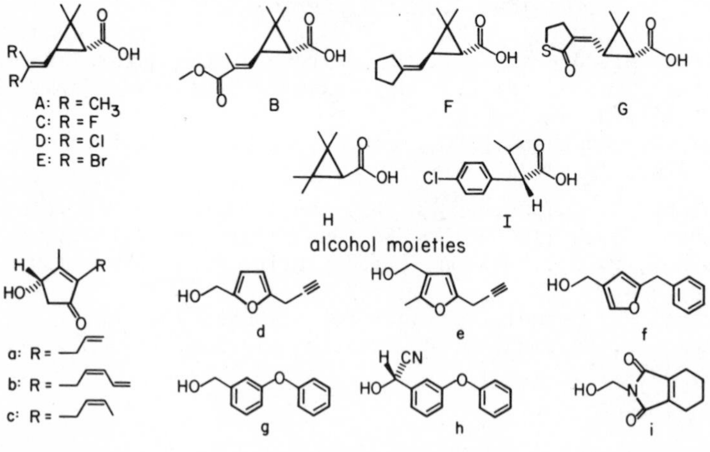

acid moieties

A: R = CH_3
C: R = F
D: R = Cl
E: R = Br

B F G H I

alcohol moieties

a: R =
b: R =
c: R =

d e f g h i

Figure 1

[1]Present address: Insecticides and Fungicides Department, Rothamsted Experimental Station, Harpenden, Hertfordshire, AL5 2JQ, England.

The most extensive information on enzymatic metabolism of pyrethroids involves studies with mouse liver microsomal enzymes (1, 4, 5). In the present investigation, the effect of structural modification on biodegradability was examined with 44 pyrethroids consisting of the above acid and alcohol moieties using mouse liver microsomal esterase and oxidase preparations (6, 7).

Methods for Enzyme Studies

Mouse hepatic microsomal preparations and their acetone powders contain esterases active in pyrethroid hydrolysis. Fresh microsomal preparations fortified with NADPH will simultaneously hydrolyze and oxidize pyrethroid substrates. To study oxidative reactions only, the fresh microsomal preparation is pretreated with an irreversible esterase inhibitor (*e.g.*, tetraethylpyrophosphate or paraoxon) and then NADPH is added (4).

Radiolabeled substrates permit identification and quantitation of individual pyrethroid metabolites thereby defining site preferences in metabolic attack by the oxidase system (5, 8). However, this approach is limited by the small number of labeled pyrethroids. The present study used unlabeled substrates to permit examination of a great variety of pyrethroids and related compounds. Metabolism rates in esterase and oxidase systems were determined as the disappearance of substrate as a function of time (4). Thus, the oxidase rates do not take into account the various sites of attack. The rates are given as derivatives of pseudo-first order constants ($K_1 \times 10^3$) where the oxidase plus esterase rate for [1R,*trans*]-resmethrin is 0.21 ± 0.05 min^{-1}.

The metabolism data are summarized in the results which follow by dividing the pyrethroid molecule into 4 regions (W-Z) using [1R,*trans*]-resmethrin as the comparison compound (Figure 2).

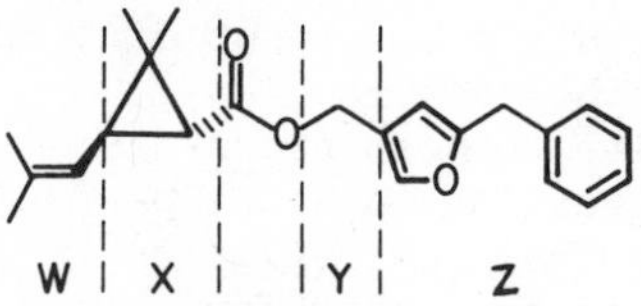

W X Y Z

Figure 2

Substrate Specificity of Pyrethroid-Hydrolyzing Enzymes

Pyrethroid hydrolysis is highly dependent on the molecular configuration, particularly in region W, the side chain at cyclopropane C-3 in the acid moiety. This is illustrated in Figure 3 with 5-benzyl-3-furylmethyl esters; a similar relationship is evident with a more limited series of 3-phenoxybenzyl esters with [1R,*trans*]- and [1R,*cis*]-acids (R'=Cl > CH_3 > Br).

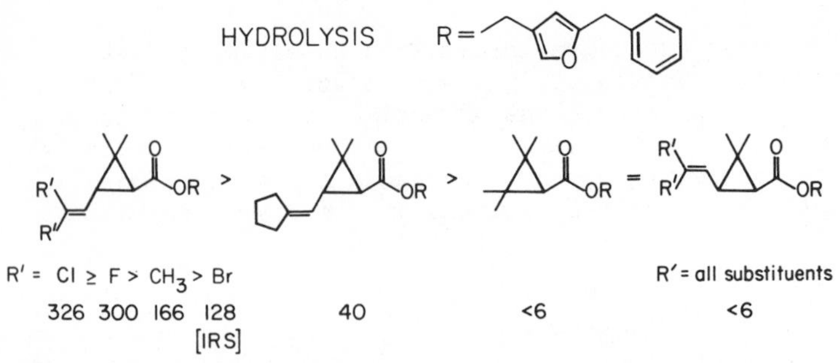

HYDROLYSIS R=

R′ = Cl ≥ F > CH_3 > Br R′= all substituents
326 300 166 128 40 <6 <6
[1RS]

Figure 3. The dependency of pyrethroid hydrolysis on molecular configuration is illustrated with 5-benzyl-3-furylmethyl esters

Only trans-substituted esters of these primary alcohols are hydrolyzed at appreciable rates; esters with a cis-substituent or a gem-dimethyl at this position are hydrolyzed poorly or not at all. The trans-substituted esters with a dichlorovinyl or difluorovinyl group are hydrolyzed most rapidly while isobutenyl and dibromovinyl esters also undergo rapid cleavage. Esters with a cyclopentylidenemethyl substituent are hydrolyzed less rapidly but are still more susceptible than cis- or gem-dimethyl-substituted esters. Thus, trans-substituents appear to be important in positioning the ester bond at the esteratic site(s).

Relatively few compounds with variations in region X were available for assay because of the importance of the gem-dimethyl group in determining insecticidal potency (9). The only structural variants in this region examined are the p-chlorophenyl-α-isopropylacetates (Ig, Ih), in which the isopropyl group substitutes for the gem-dimethyl substituent of the cyclopropane esters. The primary alcohol esters (Ig enantiomers) are slowly hydrolyzed relative to most trans-substituted cyclopropanecarboxylates, suggesting that the p-chlorophenyl group (analagous to the side chain in region W) does not properly position the remainder of the ester for rapid hydrolytic attack.

Rate differences attributable to structural variations in region X are sometimes observed between the insecticidal [1R]-cyclopropanecarboxylates and S(+)-p-chlorophenyl-α-isopropylacetates and their non-insecticidal [1S]- and R(-)-analogs; the effects of the stereochemistry at this position on both hydrolytic and oxidative metabolism are considered elsewhere (10).

Region Y, the free or substituted α-methylene position, is the most important determinant in the alcohol moiety for hydrolysis (Figure 4).

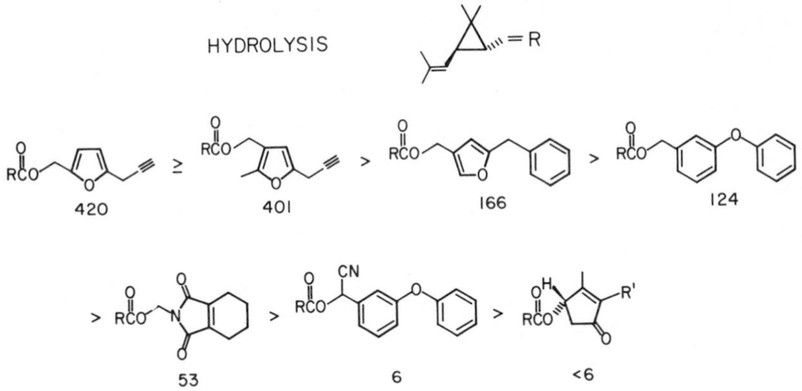

Figure 4. The free or α-methylene position (region Y) is the most important determinant in the alcohol moiety for hydrolysis

Only primary alcohol esters of appropriate acids are hydrolyzed at appreciable rates. Esters substituted at the α-position (*e.g.*, Aa, Ab, Ac, Ah, Dh and Ih) are poorly hydrolyzed or completely resistant to hydrolysis. This specificity is also seen in [1R,*trans*]-chrysanthemates of simple alcohols where the isopropyl, *sec*-butyl, *tert*-butyl and α-methylbenzyl esters are hydrolyzed at < 4-10% of the rate for the corresponding *n*-alkyl or benzyl analogs (6). Substitution at the α-position may stabilize the ester bond either directly or by interference with proper binding at the esteratic site(s).

The requirements for insecticidal activity in region Z of the alcohol moiety limited the available compounds for assay to those with a planar spacer function and a distal unsaturated center. Modifications in these components yield an 8-fold variation in hydrolysis rates for 5 primary alcohol [1R,*trans*]-chrysanthemates; furamethrin and proparthrin (Ad, Ae) are most rapidly cleaved, followed in decreasing order by resmethrin (Af), phenothrin (Ag) and tetramethrin (Ai) (Figure 4). It is therefore likely that the spacer group and unsaturated center (especially with propargylfurylmethyl compounds) of the pyrethroid alcohol moieties may cooperate with the acid side chain in positioning the molecule at the esteratic site(s).

The effect of regions Y and Z together on pyrethroid hydrolysis illustrated with [1R,*trans*]-chrysanthemates in Figure 4 is also evident with a more limited series of alcohols (f-h) esterified with the [1R,*trans*]-dichlorovinylchrysanthemic acid (D).

Substrate Specificity of Pyrethroid-Oxidizing Enzymes

The oxidase rates for pyrethroids reflect the sum of

several concurrent oxidative processes. Variations in each region therefore must be interpreted in light of the presence or absence of oxidatively-labile sites in addition to steric effects on interactions with the enzyme(s).

The oxidation rate of 5-benzyl-3-furylmethyl esters may be affected markedly by whether or not the substituent in region W undergoes oxidation but this rate is only slightly altered by the *trans*- or *cis*-configuration of this substituent (Figure 5).

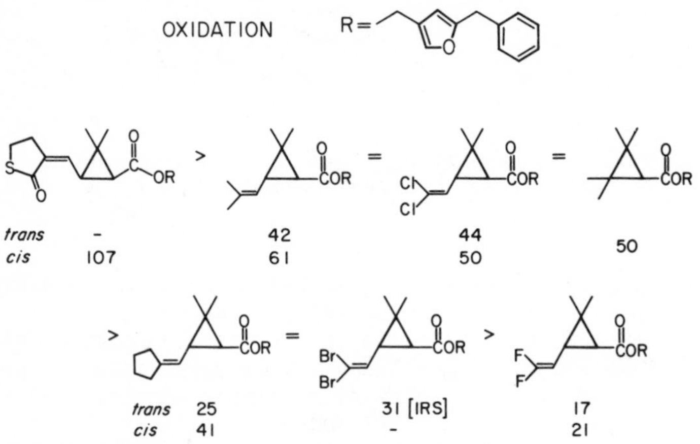

OXIDATION R=

		>		=		=	
trans	-		42		44		50
cis	107		61		50		

	>		=		>	
trans		25		31 [IRS]		17
cis		41		-		21

Figure 5

The side chain methyl groups undergo oxidase attack in allethrin (Aa; 8) and resmethrin (Af; 11) so halogen substitution at this position (*e.g.* permethrin, Dg) necessarily limits the number of hydroxylation sites in the acid moiety (12); however, substitution by a halogen to eliminate one potential site of attack does not necessarily reduce the overall susceptibility to oxidation. Esters with a dichlorovinyl group are oxidized more rapidly than either the difluorovinyl- or dibromovinyl-substituted esters, indicating that there may be an optimal substituent size at this position. The rapid oxidation with acid moiety G is probably due to sulfoxidation in the thiolactone group and the relatively low oxidase rate for ethanoresmethrin (Ff) may be due either to impaired side chain hydroxylation by the cyclopentylidenemethyl group or to an effect of this group in preventing attack in other areas of the molecule; however, these are speculative points since the sites of oxidation for these compounds are at present unknown.

The effect of alterations in region Y on oxidation rates is dependent on the nature of the substituent (Figure 6).

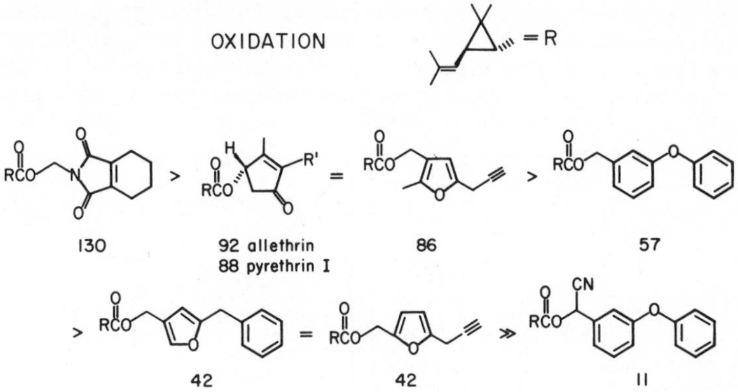

OXIDATION =R

130 > 92 allethrin, 88 pyrethrin I = 86 > 57 > 42 = 42 ≫ 11

Figure 6. The effect of alterations in region Y on oxidation rates

The cyclic secondary alcohol (e.g., a-c) esters are readily oxidized but the α-cyano-substituted (h) esters are slowly oxidized. The reduced oxidation rates for the α-cyano analogs (Ah, Dh) of phenothrin and permethrin (Ag, Dg) indicates that this modification limits oxidase attack at both the isobutenyl side chain of the acid moiety (phenothrin) and the gem-dimethyl group (permethrin). The effect on aryl hydroxylation in the alcohol moiety and the mechanism by which α-cyano substitution interferes with acid-moiety oxidation are not known.

The effect of region Z on oxidase rates is influenced by the susceptibility of sites in this region to oxidation. The alcohol moiety of tetramethrin, the most rapidly-oxidized chrysanthemate, is readily hydroxylated in the cyclohexene portion of the molecule (13). The side chains of the rethronyl and propargylfurylmethyl esters are known sites of oxidative attack (2,8). The 3-phenoxybenzyl esters are oxidized more rapidly than the corresponding 5-benzyl-3-furylmethyl esters perhaps due to increased attack in the alcohol moiety or, more likely, facilitated binding with the former compounds.

The combined effects of regions Y and Z on oxidation illustrated in Figure 6 with trans- and cis-chrysanthemates are also applicable to 3 alcohol moieties (f-h) esterified with the corresponding dichlorovinyl-substituted acids (D).

Substrate Specificity in Relation to Overall Pyrethroid Biodegradability

The data presented above on individual hydrolysis and oxidation rates for pyrethroids are applicable to considerations of overall biodegradability since their sum corresponds closely to the experimentally-determined oxidase plus esterase rates.

The pseudo-first order rate curves for [1R,trans]- and [1R,cis]-permethrin (Figure 7) illustrate the typical differences between trans- and cis-isomers: very rapid hydrolysis of the trans- relative to the cis-isomer; approximately equal oxidation

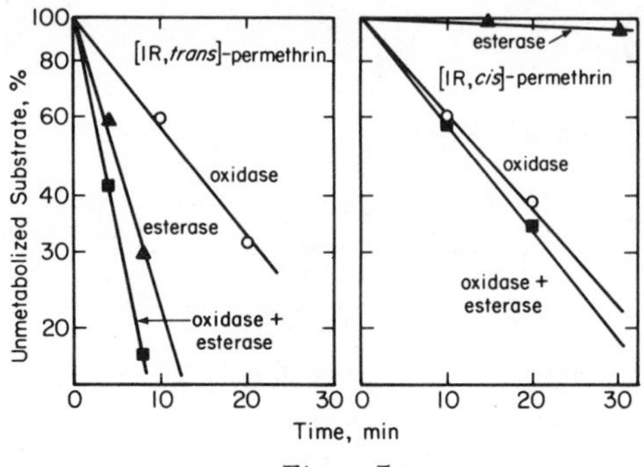

Figure 7

rates; close correspondence of the oxidase plus esterase curve to the esterase only curve in the case of the trans-isomer and to the oxidase only curve in the case of the cis-isomer; and more rapid overall (oxidase plus esterase) metabolism of the trans-isomer compared to the cis-isomer.

The overall biodegradation rate is greatest for primary alcohol esters of trans-substituted acids since they undergo rapid hydrolytic and oxidative attack (Figure 8).

BIODEGRADABILITY – ESTERASE + OXIDASE

Figure 8

Intermediate rates are found for the cis-substituted analogs, other esters with acid moieties restricting hydrolysis, and the rethrins, whose overall biodegradation depends almost entirely on oxidation. Esters of α-cyano-3-phenoxybenzyl alcohol are relatively resistant to both oxidation and hydrolysis and thus have the lowest overall esterase plus oxidase rates of the compounds examined. The acid moiety-dependent rate variations known from other series are also evident in this group, but these differences are overshadowed by the limiting effect of the α-cyano substituent.

Correlation of Microsomal Metabolism with Pyrethroid Toxicity and *in vivo* Metabolism

Studies with mice treated intraperitoneally establish that the *trans*-esters are generally less toxic than the *cis*-esters (Table I), a finding which correlates with the more rapid overall metabolism of the *trans*-isomers. In several instances, pretreatment of mice with an esterase inhibitor (DEF, Figure 9; 50 mg/kg, 6 hr prior to the pyrethroid) or an oxidase inhibitor (PB, Figure 9; 150 mg/kg, 1 hr prior to the pyrethroid) to block these enzymes *in vivo* yields significant toxicity increases thereby linking metabolic susceptibility and toxicity (Table I).

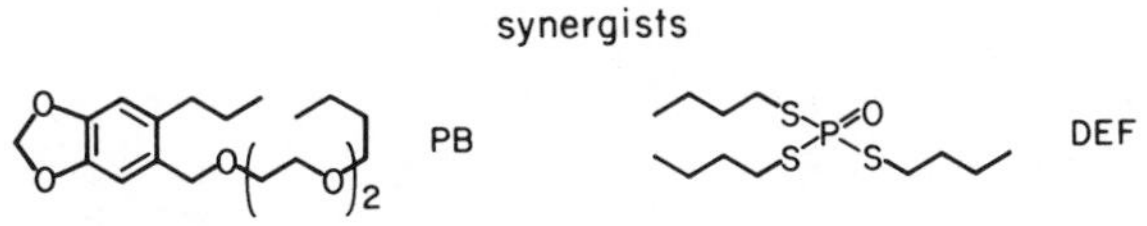

synergists

PB

DEF

Figure 9

Mice pretreated with either DEF or PB are > 10-fold more susceptible than normal mice to poisoning by *cis*-permethrin, *trans*-ethanoresmethrin, *trans*-cyanophenothrin, *trans*-NRDC 149 and S-5602. Four of the pyrethroids (*i.e.*, *trans*-resmethrin, *trans*- and *cis*-phenothrin and *trans*-permethrin) maintain their very low level of toxicity even with synergist pretreatment. All of the esters currently known to show synergist-dependent toxicity increases (Table I; see also ref. 4) also have relatively low metabolism rates in at least one microsomal enzyme system, but this is not a general rule since allethrin with a very low hydrolysis rate is not significantly synergized. Thus, it is not possible at present to predict the factor for synergism by DEF or PB from metabolism rate data with mouse liver enzymes. Detoxification enzymes other than those in hepatic microsomes and compound-dependent distribution characteristics may also influence the toxicity and synergizability.

The *in vivo* metabolism patterns of pyrethroids reflect the specificity of the pyrethroid esterase(s) determined in microsomal preparations. Primary alcohol esters of *trans*-substituted acids do not yield ester metabolites in the excreta but their *cis*-analogs give significant amounts of ester metabolites (Table II).

Table I. Mouse Intraperitoneal Toxicity and Synergism of Pyrethroids

Compound and acid configuration	Acid and alcohol moieties	LD_{50}, mg/kg trans	LD_{50}, mg/kg cis	Toxicity increase factor DEF trans	DEF cis	PB trans	PB cis	Ref.
1° alcohol esters								
Resmethrin [1R]	Af	>1500	320	none	11	none	25	4,14
Phenothrin [1R]	Ag	>1500	>1500	none	none	none	none	4,15
Permethrin [1RS]	Dg	>1000	1000	none	--[a]	none	--	15
Ethanoresmethrin [1R]	Ff	>1500	10	>188	4.2	>60	4.8	4,15
2° alcohol esters								
Allethrin [1R]	Aa[b]	30	42	1.3	1.2	1.6	--	4,15
Cyanophenothrin [1R]	Ah[c]	> 500	58	>20	3.1	>20	7	15
NRDC 149 [1RS]	Dh[c]	> 500	28	>20	5.0	>10	6	15
NRDC 161 [1R]	Eh[d]		10		2.0		3.5	15
S-3206	Hh[c,e]	15		2.2		4.1		15
S-5602 [RS]	Ih[c,e]	>500		>32		>45		15

[a]No data available; [b]trans-ester with α-S-alcohol and cis-ester with α-RS-alcohol; [c]α-RS-alcohol; [d][1R,cis]-ester with α-S-alcohol; [e]no trans/cis isomers.

Table II. Ester Metabolites of [1R,*trans*]- and [1R,*cis*]-Resmethrin and -Permethrin, Rat *in vivo*

Compound	Acid and alcohol moieties	Ester metabolites,%		Ref.
		trans	cis	
Resmethrin	Af	<0.1	9	11
Permethrin	Dg	<0.1	9	12

Esters of secondary alcohols (*i.e.*, allethrin, pyrethrin I) are much more resistant to hydrolysis *in vivo* (8). The difference in *in vivo* hydrolytic susceptibility between the *cis*-esters and the rethrins is not seen in *in vitro* studies. The observed *in vivo* hydrolysis of *cis*-esters may be due to a low level of microsomal esterase activity not measurable by the method used, to esterases in tissues or tissue fractions other than the liver microsomes, or to more efficient hydrolysis of previously-hydroxylated esters. There is some indirect evidence from *in vivo* studies on resmethrin (11) and permethrin (12) that the *cis*-isomers may hydrolyze in part as hydroxylated esters.

The substrate specificity of the mouse liver microsomal enzymes generally reflects the available *in vivo* results on mammalian toxicity and metabolite patterns. Knowledge of *in vitro* structure-biodegradability relationships should therefore be useful in interpreting *in vivo* metabolism studies and in designing pyrethroids with favorable mammalian toxicology.

Abstract

Pyrethroid structure-biodegradability relationships were examined with esterases and oxidases of mouse liver microsomes. The esterases are most important in metabolizing primary alcohol esters of cyclopropanecarboxylic acids with *trans* side chains such as isobutenyl or dihalovinyl substituents at cyclopropane C-3. The mixed-function oxidase system dominates the metabolism of all secondary alcohol esters and primary alcohol *cis*-substituted-cyclopropanecarboxylates. An α-cyano group in the alcohol greatly reduces the rate of both enzymatic hydrolysis and oxidation. Mice pretreated with an appropriate esterase or oxidase inhibitor usually show increased susceptibility to pyrethroid intoxication.

Acknowledgments

We thank Charles Abernathy, Michael Elliott, Judith Engel and Kenzo Ueda, current or former colleagues in this laboratory, for assistance and helpful suggestions. This study was supported in part by grants from: National Institutes of Health (2 P01 ES00049); Agricultural Chemical Div., FMC Corp., Middleport, N.Y.; Agricultural Chemicals Div., ICI United States Inc., Goldsboro, N. C.; McLaughlin Gormley King Co., Minneapolis, Minn.; Sumitomo Chemical Co., Osaka, Japan; Roussel-Uclaf-Procida, Paris, France; Mitchell Cotts & Co. Ltd., London, England; Wellcome Foundation Ltd., London, England; National Research Development Corp., London, England.

Literature Cited

1. Casida, J. E., Ueda, K., Gaughan, L. C., Jao, L. T., Soderlund, D. M., Arch. Environ. Contam. Toxicol. (1975/76) 3, 491.
2. Miyamoto, J., Environ. Health Perspec. (1976) 14, 15.
3. Elliott, M., Farnham, A. W., Janes, N. F., Needham, P. H., Pulman, D. A., ACS Symp. Ser. (1974) 2, 80.
4. Abernathy, C. O., Ueda, K., Engel, J. L., Gaughan, L. C., Casida, J. E., Pestic. Biochem. Physiol. (1973) 3, 300.
5. Ueda, K., Gaughan, L. C., Casida, J. E., Pestic. Biochem. Physiol. (1975) 5, 280.
6. Soderlund, D. M., Ph.D. thesis, University of California, Berkeley (1976).
7. Soderlund, D. M., Casida, J. E., Pestic. Biochem. Physiol. (1977) accepted for publication.
8. Elliott, M., Janes, N. F., Kimmel, E. C., Casida, J. E., J. Agr. Food Chem. (1972) 20, 300.
9. Elliott, M., Bull. Wld Hlth Org. (1971) 44, 315.
10. Soderlund, D. M., Casida, J. E., ACS Symp. Ser. (1977) this volume.
11. Ueda, K., Gaughan, L. C., Casida, J. E., J. Agr. Food Chem. (1975) 23, 106.
12. Gaughan, L. C., Unai, T., Casida, J. E., J. Agr. Food Chem. (1977) in press.
13. Miyamoto, J., Sato, Y., Yamamoto, K., Endo, M., Suzuki, S., Agr. Biol. Chem. (1968) 32, 628.
14. Jao, L. T., Casida, J. E., Pestic. Biochem. Physiol. (1974) 4, 456.
15. Engel, J. L., Casida, J. E., unpublished results.

16

Stereospecificity of Pyrethroid Metabolism in Mammals

DAVID M. SODERLUND[1] and JOHN E. CASIDA

Pesticide Chemistry and Toxicology Laboratory, Department of Entomological Sciences, University of California, Berkeley, Calif. 94720

For high insecticidal activity, pyrethroids must have a precise steric relationship between an unsaturated center in the alcohol moiety and the gem-dimethyl group or an equivalent substituent in the acid moiety (1). This generally requires a 1R configuration in the cyclopropanecarboxylic acid and an α-S configuration in the alcohol. Inversions at these optical centers drastically alter the potency without greatly changing the physical properties. Pyrethroid insecticides are commonly used as isomeric mixtures or, if a single isomer is involved, the residues sometimes undergo photochemical isomerization and epimerization. Metabolic studies on isomeric mixtures may not reflect the rates and sites of attack on the most bioactive components if metabolic stereoselectivity is encountered. It is therefore important to define the stereospecificity in metabolism of the optical antipodes and its relevance in pyrethroid toxicology and residue persistence.

We previously reviewed the influence of trans- and cis-substituents on the metabolism of cyclopropanecarboxylates (2-4). This report considers the stereoselectivity in in vitro and in vivo mammalian metabolism of various isomers of resmethrin, permethrin, S-5439 and S-5602 (Figure 1).

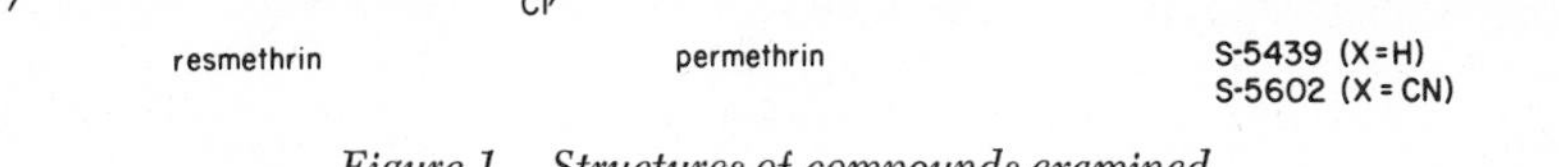

resmethrin

permethrin

S-5439 (X=H)
S-5602 (X=CN)

Figure 1. Structures of compounds examined

[1]Present address: Insecticides and Fungicides Department, Rothamsted Experimental Station, Harpenden, Hertfordshire, AL5 2JQ, England.

Resmethrin and Site Preference for Hydroxylation of Isobutenyl Methyl Groups

Mouse liver microsomes oxidize the [1R]- and [1S]-isomers of either *trans*- or *cis*-resmethrin at essentially the same rates but they hydrolyze [1R,*trans*]-resmethrin 1.7-times faster than the [1S,*trans*]-isomer (Table I). This difference in hydrolysis rate is reflected in the greater overall biodegradability (esterase plus oxidase) of [1R,*trans*]-resmethrin.

Site preference for hydroxylation of the isobutenyl methyl groups varies with both the chrysanthemate isomer and species (Figure 2).

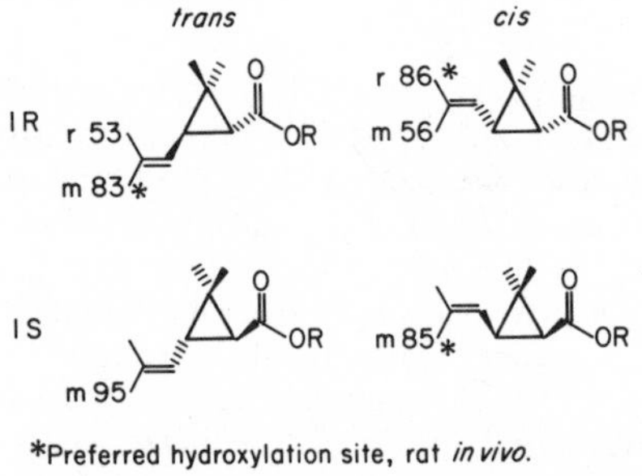

*Preferred hydroxylation site, rat *in vivo*.

Pesticide Biochemistry and Physiology

Figure 2. Stereoselectivity in hydroxylation of isobutenyl methyl groups of resmethrin isomers by mouse (m) and rat (r) microsomes (R = 5-benzyl-3-furylmethyl) (5). The percent metabolism at an indicated methyl group is relative to the sum for both methyl groups calculated by summating the identified acid-moiety metabolites.

With the mouse enzyme, hydroxylation of the *trans*(E) methyl group is preferred with both the [1R,*trans*]- and [1S,*trans*]-isomers while the *cis*(Z) methyl position is strongly preferred with the [1S,*cis*]-isomer but only slightly preferred with the [1R,*cis*]-isomer. Where data from the rat enzyme are available, the site preference is reversed from that with mouse preparations. *In vivo* data for rats (6) are consistent with the *in vitro* results for the [1R,*cis*]-isomer but not for the [1R,*trans*]-isomer.

It has been proposed (5) that an oxidative ester cleavage reaction occurs within the series of resmethrin isomers and is most important with [1S,*cis*]-resmethrin. This cleavage may result from formation of unstable hydroxylated esters by oxidation of the methylene group adjacent to the ester function or from other undefined mechanisms.

Table I. Comparative Metabolism Rates for Pyrethroid Enantiomers by Mouse Microsomal Enzymes (4)

Pyrethroid	Relative metabolism rate[a]					
	Esterase		Oxidase		Esterase + Oxidase	
	Cyclopropanecarboxylates					
	1R	1S	1R	1S	1R	1S
trans-Resmethrin	79±8	47±2	20±3	20±4	100±26	69±5
cis-Resmethrin	<3	<4	29±4	26±3	29±7	26±5
trans-Permethrin	77±4	109±9	30±5	17±4	112±2	123±4
cis-Permethrin	<2	<4	26±6	22±2	29±8	26±3
	p-Chlorophenyl-α-isopropylacetates					
	S(+)	R(-)	S(+)	R(-)	S(+)	R(-)
S-5439	<3	9±1	23±1	13±2	28±2	23±2
α-RS-S-5602	<2[b]	4±1[c]	11±3[b]	6±1[c]	15±1[b]	11±1[c]

[a]Normalized pseudo first-order rates with the esterase plus oxidase metabolism of [1R,*trans*]-resmethrin as the standard (100); [b]alcohol enantiomers are separable by glc but no differences are observed in metabolism rates; [c]alcohol enantiomers are separable by glc and metabolism rates are approximately twice as great for the first-eluting isomer.

Permethrin and Site Preference for Hydroxylation of gem-Dimethyl Group

Mouse microsomes oxidize [1R,trans]-permethrin 1.8-fold more rapidly than [1S,trans]-permethrin but there is little difference between the cis-isomers (Table I). These rate differences for oxidation of the trans-isomers are also evident with a different assay condition (substrate level 10-fold higher than the 100 nmole level involved in the experiments of Table I; 45 min incubation) which gives 51% metabolism for the [1R,trans]-isomer and 20, 37 and 34% metabolism for the [1S, trans]-, [1R,cis]- and [1S,cis]-isomers, respectively. Equivalent rat liver preparations are less active and in this case the cis-isomers are more extensively metabolized (1R,cis- 31%; 1S,cis - 27%) than the trans-isomers (1R,trans - 19%; 1S,trans - 10%).

The esterase specificity with permethrin isomers is reversed from that with resmethrin so that the [1S,trans]-isomer is hydrolyzed 1.4-fold more rapidly than the [1R,trans]-isomer (Table I). The greater hydrolysis of [1S,trans]-permethrin is sufficient to give this isomer a greater overall biodegradation rate than [1R,trans]-permethrin. The esterase: oxidase rate ratio (1R,trans=2.6; 1S,trans=6.4) illustrates the difference between the enantiomers in the relative importance of hydrolysis and hydroxylation as initial steps in metabolism.

Large amounts of 3-phenoxybenzyl alcohol, normally considered to be a hydrolysis product, are evident on glc analysis of the metabolites of [1R,trans]-permethrin but not of other permethrin isomers from mouse microsomes with oxidase but no esterase activity (pretreated with tetraethylpyrophosphate and fortified with NADPH) (3). This provided an opportunity to investigate the proposed NADPH-dependent ester cleavage reaction (see above relative to resmethrin). Mouse liver microsomes were incubated with [1R,trans]-permethrin-α-d_2 in an $^{16}O_2$ or $^{18}O_2$ atmosphere (7). Product analysis by glc-ci-ms (Table II) revealed that the 3-phenoxybenzyl alcohol retains both deuterium atoms, thereby ruling out oxidative attack at the α-methylene, a mechanism suggested for oxidative cleavage of resmethrin (5; see above). However, ^{18}O from $^{18}O_2$ was incorporated in the γ-lactone as evidenced by ms comparison (Table II) of this product from both $^{16}O_2$ and $^{18}O_2$ incubations with the authentic lactone (8). The propensity of the 2-cis-hydroxymethyl carboxylic acids to form γ-lactones under acidic conditions (8) or on heating led to the hypothesis that an unstable hydroxy ester was formed and then degraded (Figure 3). The instability of the authentic hydroxy ester was verified by the finding of quantitative formation of lactone and alcohol on glc. This artifact of analysis is overcome by treating the microsomal extracts with N,O-bis(trimethylsilyl)acetamide (BSA) to form the trimethylsilyl (TMS) derivative, 2-cis-TMSO-[1R,trans]-

Table II. Mass Spectra of [1R,trans]-Permethrin, [1R,trans]-Permethrin-α-d_2 and Their Mouse Microsomal Oxidase Metabolites and Derivatives (7)

Compound (source[a])	Glc Rt,[b] min	Ci-ms (isobutane), m/e (rel. intensity)[c]		
		$[M+1]^+$	$[M-Cl]^+$	$[3\text{-PhOPhCH}_2]^+$ or $[3\text{-PhOPhCD}_2]^+$
[1R,trans]-Permethrin (std)	12.8	391(35)	355(47)	183(100)
[1R,trans]-Permethrin-α-d_2 (std)	12.8	393(30)	357(56)	185(100)
3-PhOPhCH$_2$OH (std)	6.4	201(10)		183(100)
3-PhOPhCD$_2$OH ($^{16}O_2$ and $^{18}O_2$)	6.4	203(9)		185(100)
γ-Lactone (std and $^{16}O_2$)	4.1	207(100)	171(6)	
γ-Lactone-^{18}O ($^{18}O_2$)	4.1	209(100)	173(5)	
2-cis-TMSO-[1RS,trans]-Permethrin (std)	14.2	479(95)		183(23)
2-cis-TMSO-[1R,trans]-Permethrin-α-d_2 ($^{16}O_2$)	14.2	481(91)		185(19)
2-cis-TMS^{18}O-[1R,trans]-Permethrin-α-d_2 ($^{18}O_2$)	14.2	483(97)		185(20)

[a] Sources--std = standard from synthesis, $^{16}O_2$ and $^{18}O_2$ = gas phase for microsomal reaction; [b] 2 m x 2 mm i.d. OV-101 (3.8%) on Chromosorb W (60-80 mesh), temperature program 150-300°C at 10°C/min; [c] assignments given only for the most abundant isotopes in the ion clusters.

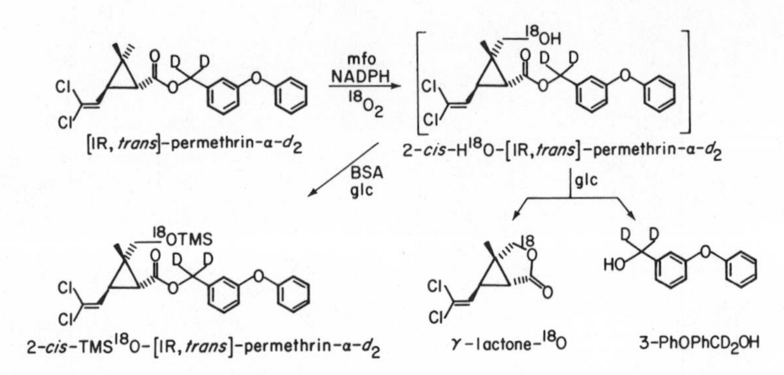

Figure 3. Metabolism of dideuteropermethrin by mouse microsomes in an $^{18}O_2$ atmosphere

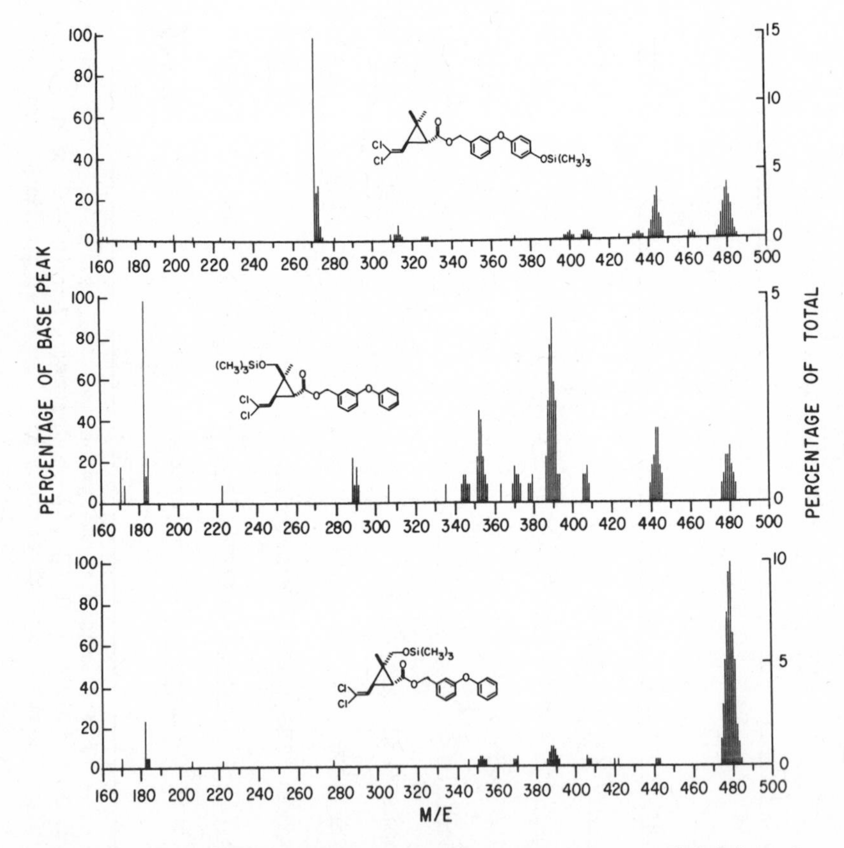

*Figure 4. CI-MS (isobutane) fragmentation patterns for trimethylsilyloxy derivatives of [1R,*trans*]-permethrin*

permethrin-α-d_2, identified by glc-ci-ms comparison with the authentic BSA-derivatized non-deuterated compound (Table II, Figure 4).

The TMS derivatization technique was used for quantitation (glc-flame ionization detector) and identification (glc-ci-ms) of metabolites of the [1R]- and [1S]-isomers of *trans*- and *cis*-permethrin by comparison with the TMS derivatives of the authentic hydroxy-[1RS]-permethrin isomers (8). One of the compounds, 2-*cis*-TMSO-[1R,*cis*]-permethrin, was available only as a metabolite and not as an authentic [1RS]-standard. Using glc conditions given in Table II, the 4'-TMSO-, 2-*trans*-TMSO- and 2-*cis*-TMSO-derivatives of *trans*-permethrin give Rt values of 14.9, 14.4 and 14.2 min, respectively; the corresponding derivatives of *cis*-permethrin give Rt values of 14.8, 14.4 and 14.0 min, respectively. Each of the TMS-derivatives of the hydroxy-*trans*-permethrin isomers gives a distinctive ci-ms fragmentation pattern (Figure 4). The fragmentation pattern of 4'-TMSO-*trans*-permethrin (upper spectrum) is characterized by the 4'-trimethylsilyloxy-3-phenoxybenzyl cation (*m/e* 271) and smaller clusters for the quasi-molecular ion ($[M+1]^+$, *m/e* 479) and for this ion minus HCl ($[M-Cl]^+$, *m/e* 444). 2-*trans*-TMSO-*trans*-permethrin (middle spectrum) is identified by the 3-phenoxybenzyl cation (*m/e* 183) and loss of the trimethylsilyloxy function ($[M-OTMS]^+$, *m/e* 389) with smaller ion clusters corresponding to $[M+1]^+$ and $[M-Cl]^+$. 2-*cis*-TMSO-*trans*-permethrin (lower spectrum) gives a large quasi-molecular ion cluster ($[M+1]^+$, *m/e* 479) and little further fragmentation. Mass spectra of the corresponding TMSO-*cis*-permethrin isomers revealed the same general fragmentation patterns as their *trans*-permethrin analogs with minor ion intensity differences.

Site preference for hydroxylation of the 4 permethrin isomers by mouse and rat microsomes was determined as above with TMS derivatives using intercomparison of several ci-ms spectra through a glc peak where necessary to eliminate the possibility of the presence of a second minor component in a metabolite peak (*i.e.*, 2-*trans*- and 2-*cis*-TMSO-*trans*-permethrin). Since epimerization of permethrin isomers at either cyclopropane C-1 or C-3 is not a known metabolic reaction *in vivo* (9) the configuration of the metabolites at C-1 and C-3 is assumed to be the same as that of the parent ester.

The isomer-dependent site preference for hydroxylation of permethrin is shown in Figure 5. With the *trans*-isomers, both mouse and rat microsomes carry out stereospecific oxidation of one methyl group, hydroxylating the 2-*cis*-methyl of [1R,*trans*]-permethrin and the 2-*trans*-methyl of [1S,*trans*]-permethrin. Aryl hydroxylation at the 4'-phenoxy position of the *trans*-isomers is significant only with the rat oxidase. Methyl group oxidation is less specific with the *cis*-esters; the mouse enzyme preferentially hydroxylates the 2-*cis*-position of [1R,*cis*]-permethrin and the 2-*trans*-position of [1S,*cis*]-permethrin but

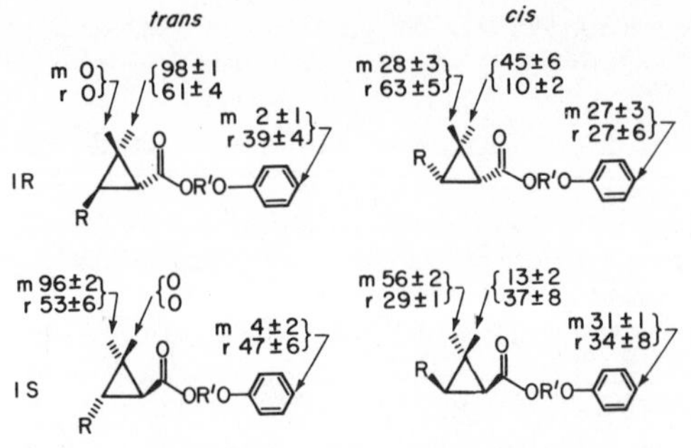

Journal of Agricultural and Food Chemistry

Figure 5. Stereoselectivity in hydroxylation of gem-*dimethyl group and phenoxy ring of permethrin isomers by mouse (m) and rat (r) microsomes (R = 2,2-dichlorovinyl; R′ = 3-substituted-benzyl). The percent metabolism at an indicated site is relative to the sum for all sites.*

with the rat enzyme this pattern is reversed. Hydroxylation at the 4'-position is a major pathway with both cis-isomers in both enzyme systems.

Studies on in vivo metabolism of the trans- and cis-isomers of [1R]- and [1RS]-permethrin in rats provide data for comparing in vivo (Figure 6) and in vitro (Figure 5) specificities.

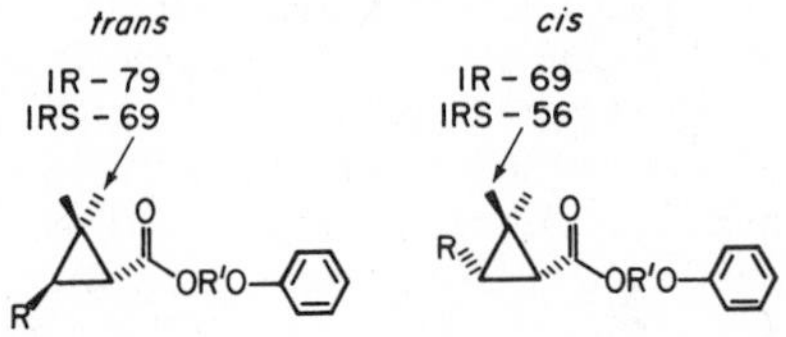

Figure 6. Stereoselectivity in hydroxylation of gem-*dimethyl group of permethrin isomers by rats in vivo (R = 2,2-dichlorovinyl; R′ = 3-substituted-benzyl) (9). The percent metabolism at an indicated methyl group is relative to the sum for both methyl groups calculated by summating identified acid-moiety metabolites.*

The [1R]-isomers show clear methyl group preference in vivo similar to that found in vitro and, as expected, there is less overall specificity with the [1RS]-isomers. The in vivo ester cleavage reaction complicates these comparisons since oxidation of the hydrolysis products may involve a different preference in hydroxylation site than with the esters themselves. Thus, the lack of methyl group preference in in vivo hydroxylation of the acid moiety of [1R,trans]-permethrin (9) may explain the differences in specificity between in vivo and in vitro studies.

p-Chlorophenyl-α-isopropylacetates (S-5439 and S-5602)

The most striking enantiomer-dependent differences in metabolism rates are with the S(+)- and R(-)-isomers of S-5439 and S-5602 (Table I, Figure 7).

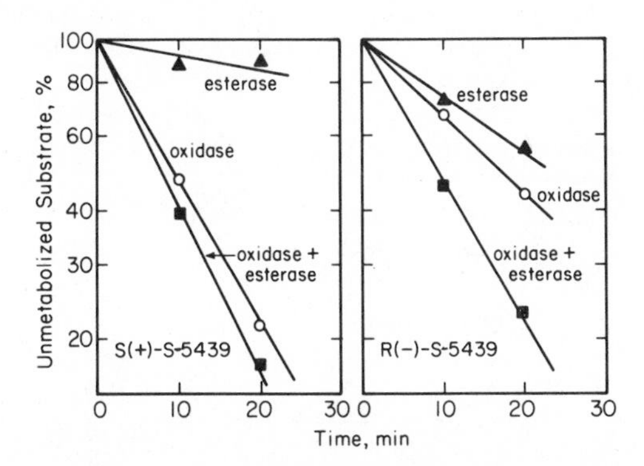

Figure 7. Metabolic rates for S-5439 isomers in mouse microsomal systems (7)

The insecticidal S(+)-S-5439 is hydrolyzed very slowly while the non-insecticidal R(-)-isomer is hydrolyzed at a moderate rate. However, oxidation of the S(+)-isomer proceeds twice as rapidly as that of the R(-)-isomer so the resulting total biodegradation rates (esterase plus oxidase) are very similar. These relationships also hold for S-5602 but the rates are only about half those of the corresponding S-5439 isomers (Table I).

The R- and S-α-cyano enantiomers of each S-5602 isomer are partially separable by glc so that differences in metabolism rates can be measured although the absolute configurations are not assigned. With R(-)-S-5602, one of the α-cyano enantiomers is both hydrolyzed and oxidized about twice as rapidly as the other, a difference not evident with the enantiomers of S(+)-S-5602 (Table I).

Enantiomer and species differences in microsomal oxidation of S-5439 were further investigated by the TMS derivatization and glc and glc-ci-ms techniques used for the permethrin study. Only one S-5439 metabolite peak was detected after TMS derivatization in studies on both the S(+)- and R(-)-isomers with either mouse or rat microsomes. Larger amounts of this metabolite were formed from the S(+)- than from the R(-)-isomer in both the rat and mouse systems, a finding consistent with available oxidase rate data for these enantiomers (Table I).

This metabolite is tentatively identified by isobutane and methane glc-ci-ms with and without BSA derivatization as 3-phenoxybenzyl 2-(4'-chlorophenyl)-3-methyl-4-hydroxybutyrate (Table III). Isobutane spectra of the TMS derivative (TMSO-S-5439) show the quasi-molecular ion cluster as the base peak while methane spectra give the 3-phenoxybenzyl cation as the base peak, establishing that hydroxylation occurs in the acid rather than the alcohol moiety. When the metabolite is analyzed by glc without derivatization, it gives a peak for 3-phenoxybenzyl alcohol plus 2 overlapping peaks at short retention times with identical methane CI spectra. The short retention time peaks are probably the 2 diastereomers of the γ-lactone, i.e. hydroxylation of a methyl group introduces a second chiral center.

The major oxidation reaction for both isomers of S-5439 in mouse and rat enzymes appears to involve hydroxylation of a methyl group analogous to the observed gem-dimethyl hydroxylation with permethrin. This hydroxylated ester is probably unstable on glc and cleaves to the corresponding γ-lactone and alcohol if not derivatized (Figure 8).

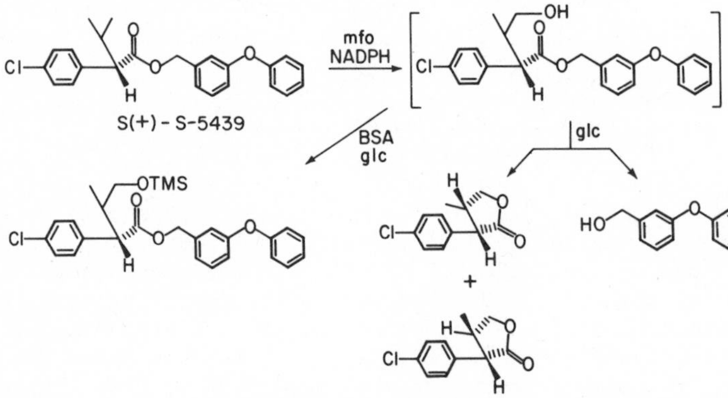

Figure 8. Proposed metabolism of S-5439 by mouse and rat microsomes

It is surprising to find that only the methyl position is hydroxylated in S-5439 since the tertiary-position of the isopropyl group and the 4-position of the phenoxy group are also potential sites for hydroxylation. However, no metabolites with glc or glc-ci-ms properties appropriate for these compounds are detected on mouse or rat microsomal metabolism of S-5439.

Discussion

Pyrethroid hydrolysis by mouse microsomal esterases is greatly facilitated by a trans-isobutenyl or equivalent substituent at cyclopropane C-3 but is affected to a much lesser

Table III. Mass Spectra of the Trimethylsilyloxy-Derivative and Degradation Products of a S-5439 Mouse and Rat Microsomal Oxidase Metabolite

Compound (ci reagent gas)	Glc Rt, min[a]	m/e (Rel. intensity)[b]	
		$[M+1]^+$	$[PhOPhCH_2]^+$
Metabolite as TMS derivative			
TMSO-S-5439 (isobutane)	15.2	483(100)	183(5)
TMSO-S-5439 (methane)	15.2	483(60)	183(100)
Products from glc of underivatized metabolite			
γ-Lactone (methane)	5.8[c]	211(100)	
3-Phenoxybenzyl alcohol (methane)	6.4	201(34)	183(100)

[a] Conditions same as for TMSO-derivatives of permethrin (footnote b, Table II); [b] assignments given only for most abundant isotopes in the ion clusters; [c] partially resolved into 2 peaks (5.0 and 5.2 min) with glc conditions as in Table II (footnote b) but isothermal at 175°C, each component giving the same mass spectrum.

degree by the optical configuration at cyclopropane C-1 (3,4, 10). The p-chlorophenyl-α-isopropylacetates, with only one optical center in the acid moiety, show greater enantiomer-dependent differences in hydrolysis rates.

Variable degrees of stereospecificity are involved in pyrethroid hydroxylation by mouse microsomal oxidases. With resmethrin the geometrical configuration of the acid (trans or cis) is most important in determining the preference between the isobutenyl methyl groups while [1R]- and [1S]-enantiomer differences are minor. The preferred site for hydroxylation of permethrin isomers is dependent on the configuration at C-1; thus, the [1R]-isomers show a preference for 2-cis-methyl hydroxylation and the [1S]-isomers for 2-trans-methyl hydroxylation. With the trans-permethrin isomers the stereospecificity of gem-dimethyl hydroxylation is absolute and there are marked rate differences between enantiomers. There are also significant rate differences in methyl hydroxylation of the S-5439 enantiomers. Mouse and rat microsomes sometimes differ in site preference for methyl hydroxylation (i.e., isobutenyl groups of [1R,trans]- and [1R,cis]-resmethrin and gem-dimethyl groups of [1R,cis]- and [1S,cis]-permethrin) and in the significance of aryl hydroxylation (i.e., [1R,trans]- and [1S,trans]-permethrin).

In studies on pyrethroid metabolism, toxicology and residues, it must be recognized that the extent and significance of metabolic stereospecificity may vary with different esters and their optical antipodes and with different species.

Abstract

Metabolism rates of the optical antipodes of pyrethroids by mouse liver microsomal enzymes differ by at least 1.7-fold in the following cases: hydrolysis of [1R,trans]- vs [1S,trans]-resmethrin; hydrolysis of S(+)- vs R(-)-S-5439 and -S-5602; methyl hydroxylation of [1R,trans]- vs [1S,trans]-permethrin; methyl hydroxylation of S(+)- vs R(-)-S-5439; oxidation of S(+)- vs R(-)-S-5602; hydrolysis and oxidation of αR- vs αS-R(-)-S-5602. Various degrees of stereospecificity are encountered with mouse and rat liver microsomal oxidases in the preferred methyl group for hydroxylation, i.e., trans(E) vs cis(Z) in the isobutenyl moiety of the 4 resmethrin isomers and 2-trans vs 2-cis in the 4 permethrin isomers. Stereospecificity in methyl hydroxylation of [1R]- and [1RS]-preparations of trans- and cis-permethrin is also evident with rats in vivo.

Acknowledgements

We thank Michael Elliott, Loretta Gaughan, Roy Holmstead, Kenzo Ueda and Tadaaki Unai, current or former colleagues in this laboratory, for assistance and helpful suggestions. This study was supported in part by grants from: National Institutes of Health (2P01 ES00049); Agricultural Chemical Div., FMC Corp., Middleport, N.Y.; Agricultural Chemicals Div., ICI United States Inc., Goldsboro, N.C.; McLaughlin Gormley King Co., Minneapolis, Minn.; Sumitomo Chemical Co., Osaka, Japan; Roussel-Uclaf-Procida, Paris, France; Mitchell Cotts & Co. Ltd., London, England; Wellcome Foundation Ltd., London, England; National Research Development Corp., London, England.

Literature Cited

1. Elliott, M., Farnham, A. W., Janes, N. F., Needham, P. H., Pulman, D. A., ACS Symp. Ser. (1974) 2, 80.
2. Casida, J. E., Ueda, K., Gaughan, L. C., Jao, L. T., Soderlund, D. M., Arch. Environ. Contam. Toxicol. (1975/76) 3, 491.
3. Soderlund, D. M., Casida, J. E., ACS Symp. Ser. (1977) this volume.
4. Soderlund, D. M., Casida, J. E., Pestic. Biochem. Physiol. (1977) accepted for publication.
5. Ueda, K., Gaughan, L. C., Casida, J. E., Pestic. Biochem. Physiol. (1975) 5, 280.
6. Ueda, K., Gaughan, L. C., Casida, J. E., J. Agr. Food Chem. (1975) 23, 106.
7. Soderlund, D. M., Ph.D. thesis, University of California, Berkeley, 1976.
8. Unai, T., Casida, J. E., ACS Symp. Ser. (1977) this volume.
9. Gaughan, L. C., Unai, T., Casida, J. E., J. Agr. Food Chem. (1977) in press.
10. Miyamoto, J., Suzuki, T., Nakae, C., Pestic. Biochem. Physiol. (1974) 4, 438.

17

Permethrin Metabolism in Rats and Cows and in Bean and Cotton Plants

LORETTA C. GAUGHAN, TADAAKI UNAI, and JOHN E. CASIDA

Pesticide Chemistry and Toxicology Laboratory, Department of Entomological Sciences, University of California, Berkeley, Calif. 94720

The discovery that 3-phenoxybenzyl 3-(2,2-dichlorovinyl)-2,2-dimethylcyclopropanecarboxylate (permethrin) combines outstanding insecticidal activity, low mammalian toxicity and adequate stability in light and air has focused attention on the potential of synthetic pyrethroids in agricultural pest insect control (1,2). Most permethrin preparations are [1RS, trans,cis]-mixtures, the [1R,trans]- and [1R,cis]-isomers being the insecticidal components (3). The importance of understanding permethrin biodegradation prompted the present study on the comparative metabolism of [1R,trans]-, [1RS, trans]-, [1R,cis]- and [1RS,cis]-permethrin in rats (4,5) and of [1RS,trans]- and [1RS,cis]-permethrin in cows and in bean and cotton plants.

[^{14}C]Permethrin Preparations and Experimental Procedures for Rats and Cows

Eight [^{14}C]permethrin preparations were used with specific activities ranging from 1.7 to 58.2 mCi/mmole (Figure 1). The [1R]-isomers were prepared as previously reported (4) and the [1RS]-isomers were provided by FMC Corporation (Middleport, N.Y.). Rats (male, albino, Sprague-Dawley strain) treated with a

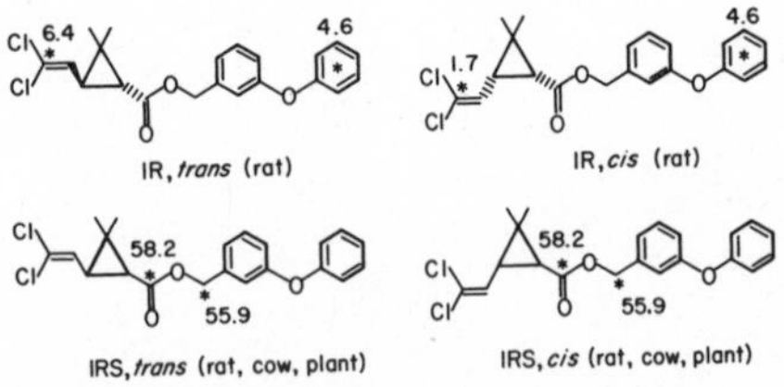

IR, *trans* (rat) — 6.4, 4.6

IR, *cis* (rat) — 1.7, 4.6

IRS, *trans* (rat, cow, plant) — 58.2, 55.9

IRS, *cis* (rat, cow, plant) — 58.2, 55.9

Figure 1. Eight preparations of ^{14}C-permethrin (specific activity, mCi/mmol)

single oral dose of each of the 8 labeled preparations at 1-4 mg/kg were held 4- or 14-days in metabolism cages and then sacrificed for radioanalysis of urine, feces, CO_2 and tissues. Metabolites in the urine (40-100 µl) and in the feces (methanol extract equivalent to 40-230 mg feces) were subjected to tlc cochromatography on silica gel chromatoplates with standard compounds (6) or their methylated (CH_2N_2) derivatives or with glucuronides synthesized enzymatically (5). Individual metabolites isolated by tlc were also hydrolyzed with enzymes (β-glucuronidase, aryl sulfatase) or with acid and base to obtain cleavage products for tlc cochromatographic identification with and without derivatization. Urinary metabolites were chromatographed in acidic solvent systems to move all products free from the origin. The fecal metabolites were chromatographed in both acidic and neutral solvent systems, the latter to minimize decomposition of ester metabolites.

The studies comparing all 8 [^{14}C]preparations in rats gave very similar results for the Cl_2C^*= and -C^*(O)- labels in the acid moiety and for the -C^*H_2- and phenoxy* labels in the alcohol moiety and with no $^{14}CO_2$ production in any case. These findings indicate that either of the [^{14}C]acid preparations or [^{14}C]alcohol preparations can be used to detect all of the metabolites from the acid and alcohol moieties, respectively. In addition, the [1R]- and [1RS]-isomers gave almost identical results, so the [1RS]-isomers with specific activities of 55.9-58.2 mCi/mmole were used in the studies with cows and other organisms.

Cows (lactating Jersey, arrangements by FMC Corporation) treated with 3 consecutive daily doses by intubation into the rumen of the 4 labeled preparations of [1RS]-permethrin at 1 mg/kg were held 12- or 14-days prior to sacrifice and analyses as above.

[^{14}C]Permethrin Metabolites in Rats and Cows

The [1RS,*trans*]-isomer of permethrin yields more urinary radiocarbon than [1RS,*cis*]-permethrin with either acid- or alcohol-labeled preparations and with either rats or cows (Table I).

Table I. Percent Urinary Radiocarbon from [1RS,*trans*]- and [1RS,*cis*]- [^{14}C]Permethrin Preparations

Isomer and label position	Rats	Cows
1RS,*trans*		
Acid	82	39
Alcohol	79	47
1RS,*cis*		
Acid	54	29
Alcohol	52	22

The majority of the metabolites appear in the urine with rats and in the feces with cows. These results indicate more extensive ester cleavage or conjugation of the metabolites with trans-permethrin than with cis-permethrin and in rats as compared to cows.

The tlc cochromatographic technique for metabolite identification is illustrated in Figure 2 with the metabolites from the acid moiety of [1RS,trans]-permethrin in rats and cows.

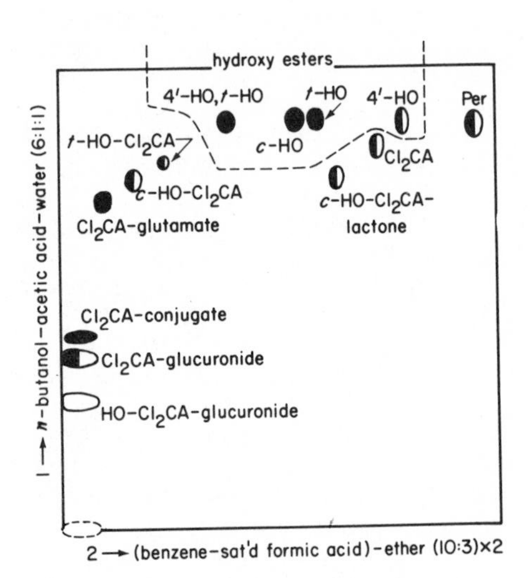

Figure 2. Metabolites from acid moiety of (IRS, trans*)-permethrin.* ○ *represents rat;* ●, *cow; and* ◑, *rat and cow.*

The solvent system for the first direction of development separates the conjugates and the second solvent system then resolves permethrin, its derivatives with monohydroxylation at the trans- or cis-position of the gem-dimethyl group relative to the carboxyl group (t-HO; c-HO), its 4'-HO derivative from phenoxy hydroxylation, its dihydroxy derivative (4'-HO,t-HO), the acid moiety (Cl_2CA) and its hydroxy derivatives (t-HO-Cl_2CA; c-HO-Cl_2CA), and the lactone of c-HO-Cl_2CA (from cyclization before excretion or as an artifact from cyclization on analysis). Most of the metabolites are formed by both rats and cows. However, only cows give ester metabolites hydroxylated at the gem-dimethyl group, the glutamate conjugate of Cl_2CA and an additional unidentified metabolite of the acid moiety. In contrast, only rats form glucuronides of the HO-Cl_2CA derivatives.

Studies of the type indicated above with each labeled preparation of [1RS,trans]- and [1RS,cis]-permethrin served to define the sites of metabolic attack in rats and cows (Figure 3). There are 4 principal sites of attack in each case, with an additional site for rats administered [1RS,cis]-

Ester hydrolysis (1):
trans ≫ *cis*
Preferred methyl for hydroxylation (2, 3):
cis (2) in 1RS, *trans*
trans (3) in 1RS, *cis*
Phenoxy hydroxylation (4, 5):
4′ in rat and cow
2′ in rat, *cis* only

Figure 3. Sites of metabolic attack from rat and cow

permethrin. Ester hydrolysis (1) is more rapid with trans- than with cis-permethrin. Oxidation at the gem-dimethyl group occurs selectively at the cis-position (2) in [1RS,trans]-permethrin and at the trans-position (3) in [1RS,cis]-permethrin. The phenoxy group is hydroxylated at the 4'-position (4) with both isomers in rats and cows and at the 2'-position (5) with cis-permethrin in rats only.

Eight ester metabolites hydroxylated in the acid or alcohol moiety are identified from the feces of rats and cows (Figure 4). Three of the 4 possible esters from monohydroxylation at the gem-dimethyl group appear in cow feces but only the 2-trans-hydroxy compound from the more metabolically-stable cis-permethrin isomer appears in rat feces. The 4'-hydroxy derivative is present with both trans- and cis-

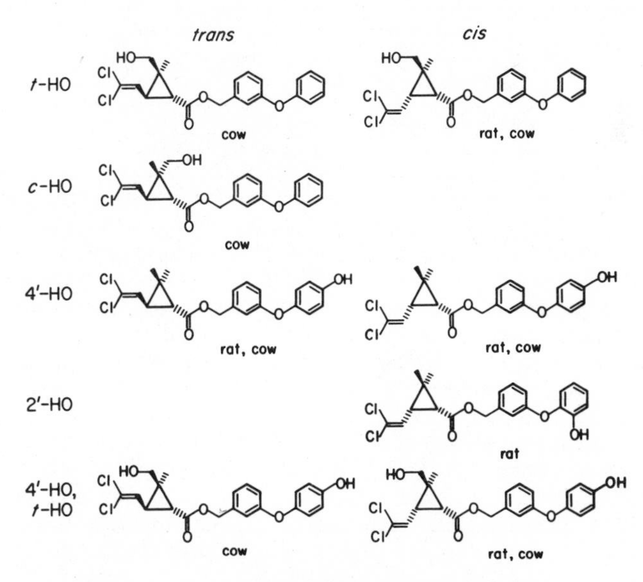

Figure 4. Rat and cow hydroxy ester metabolites

permethrin and in both rats and cows. The 4'-HO,*t*-HO-derivative appears only in cows with *trans*-permethrin and in both rats and cows with *cis*-permethrin. The feces of rats administered *cis*-permethrin contains the 2'-hydroxy derivative. These species differences are attributable in part to less extensive ester hydrolysis in cows than in rats and the ability of rats to carry out aryl hydroxylation at the 2'-position.

The acid moieties from [1RS]-*trans*- and -*cis*-permethrin are mostly excreted in rats and cows as the corresponding glucuronides. The other metabolites are also the same in both species except that in cows the glucuronides of the hydroxy acids are not detected and Cl_2CA is conjugated in part with glutamic acid (Figure 5).

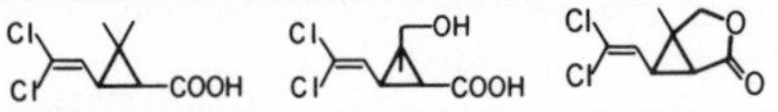

Species	free	gluc	free	gluc	
Rat	+	++	+	+	+
Cow	+	++*	+	-	+

*also glutamate conjugate of *trans*-acid

Figure 5. Metabolites from acid moiety of (IRS, trans*)-permethrin and (IRS,*trans*)-permethrin, rat and cow*

The alcohol moiety liberated on cleavage of [1RS,*trans*]- and [1RS,*cis*]-permethrin is in the most part further oxidized to the corresponding benzoic acid which is excreted free in rats, as a glycine conjugate and glucuronide in rats and cows and as the glutamate conjugate which is the major metabolite in cows but absent in rats (Figure 6). 3-Phenoxybenzyl

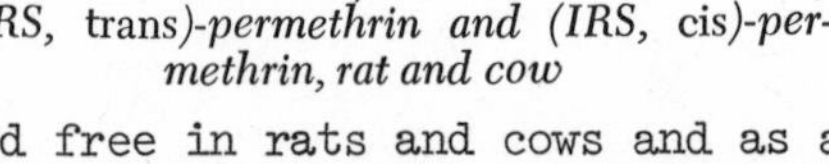

Species	free	gly	gluc	glut	free	gluc	4'	2'
Rat	+	+	+	-	+	-	++	+
Cow	-	+	+	++	+	+	+	-

Figure 6. Metabolites from the alcohol moiety of (IRS, trans*)-permethrin and (IRS,* cis*)-permethrin, rat and cow*

alcohol is excreted free in rats and cows and as a trace amount of glucuronide in cows only. The major rat metabolite, the sulfate of the 4'-hydroxy acid, is present in small amount in cow urine and the sulfate of the 2'-hydroxy derivative appears only in rat urine.

The complete metabolic pathway for *trans*- and *cis*-permethrin in rats including the 24 identified metabolites (5) is shown in Figure 7. This pathway accounts for all permethrin metabolites excreted in amounts of >1% of the administered radiocarbon except for 5 minor fecal metabolites of *cis*-permethrin.

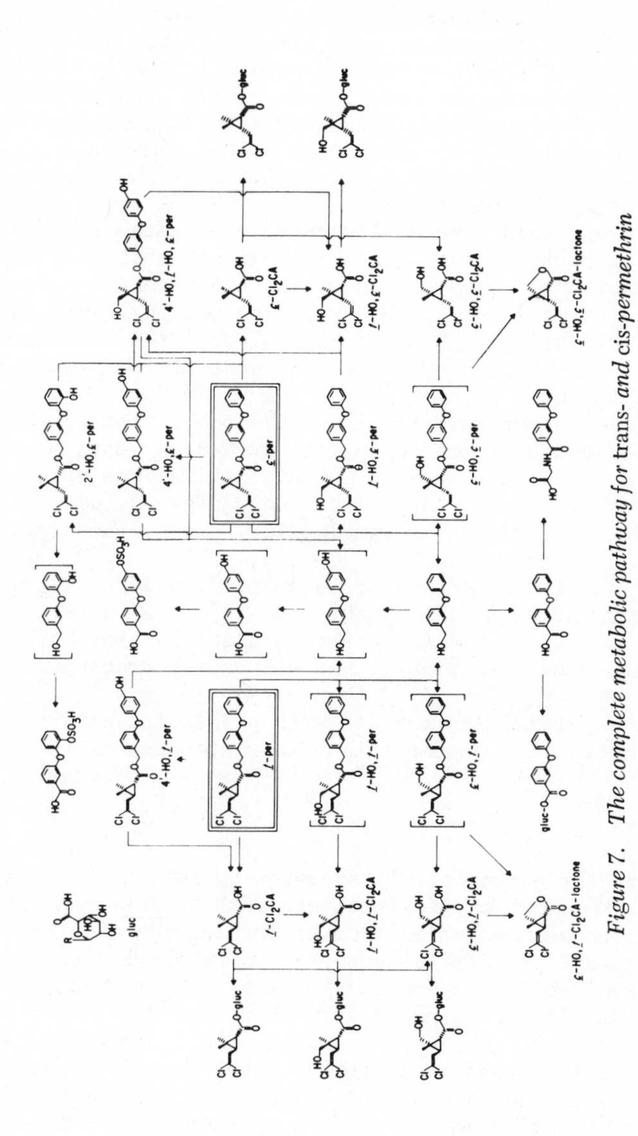

Figure 7. The complete metabolic pathway for trans- *and* cis-*permethrin*

Experimental Procedures for Bean and Cotton Plants and [^{14}C]-Permethrin Metabolites in Plants

Bean and cotton plants treated topically on the leaves and bean plants treated by stem injection with the 4 [^{14}C]-preparations of [1RS]-permethrin were held for up to 21 days in the greenhouse. The leaves were washed with methanol-chloroform (2:1) and then extracted with this solvent mixture or the whole plants were extracted in the same manner. Plant metabolites were identified and analyzed by the methods used for the mammalian metabolites, except that β-glucosidase, cellulase and acid were employed for conjugate cleavage.

An extract of bean plants 14 days after stem injection with [1RS,*trans*]-permethrin contains the parent compound, hydroxylated permethrin, the free dichlorovinyl acid and its hydroxy derivatives, phenoxybenzyl alcohol and phenoxybenzoic acid. These products appear as one spot in the *n*-butanol-acetic acid-water solvent system (Figure 8) but they are resolved in the benzene(formic acid)-ether system. The products at Rf 0.47 and 0.61 are conjugates of hydroxylated permethrin while the Rf 0.56 product is an unidentified conjugate from the acid moiety. The identified conjugates include the glycosides of the dichlorovinyl acid and of 3-phenoxybenzyl alcohol.

Permethrin on bean and cotton leaves undergoes *trans-cis* isomerization to the extent of 6-13% in 21 days. The penetrated portion yields metabolites similar to those found in the injected bean plants. In all cases, *trans*-permethrin is more rapidly metabolized than *cis*-permethrin.

These preliminary results with plants indicate the importance of photodecomposition and metabolic oxidation and hydrolysis in the dissipation of permethrin residues.

Abstract

Permethrin metabolites excreted by rats and cows include 8 mono- and dihydroxy derivatives of the *trans*- and *cis*-esters, the acid moieties from ester cleavage and their 2-*trans*- and 2-*cis*-hydroxy derivatives, 3-phenoxybenzyl alcohol, and 3-phenoxybenzoic acid and its 2'- and 4'-hydroxy derivatives. These metabolites are excreted without conjugation or as glucuronides and glycine and glutamic acid conjugates of the carboxylic acids and as sulfates of the phenolic compounds. Permethrin on bean and cotton leaves undergoes *trans-cis*-photoisomerization and the absorbed material yields hydroxy esters and their glycosides, hydrolysis products and their glycosides, and 3-phenoxybenzoic acid. *trans*-Permethrin generally undergoes more rapid biodegradation than *cis*-permethrin, in part because of the greater hydrolysis rate of the *trans*-compound.

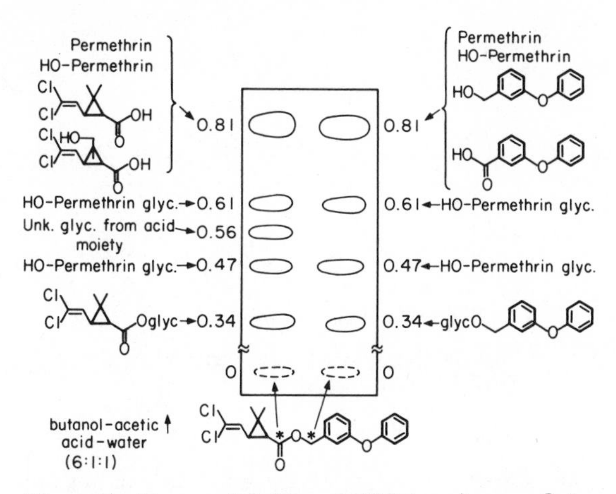

Figure 8. Bean metabolites of (IRS, trans*)-permethrin, stem injection*

Acknowledgments

We thank Michael Elliott for assistance and helpful suggestions. This study was supported in part by grants from: National Institutes of Health (2 P01 ES00049); Agricultural Chemical Div., FMC Corp., Middleport, N.Y.; Agricultural Chemicals Div., ICI United States Inc., Goldsboro, N. C.; Sumitomo Chemical Co., Osaka, Japan; Roussel-Uclaf-Procida, Paris, France; Mitchell Cotts & Co. Ltd., London, England; Wellcome Foundation Ltd., London, England; National Research Development Corp., London, England.

Literature Cited

1. Elliott, M., ACS Symp. Ser. (1977) this volume.
2. Elliott, M., Farnham, A. W., Janes, N. F., Needham, P. H., Pulman, D. A., Stevenson, J. H., Nature (1973) 246, 169.
3. Elliott, M., Farnham, A. W., Janes, N. F., Needham, P. H., Pulman, D. A., Pestic. Sci. (1975) 6, 537.
4. Elliott, M., Janes, N. F., Pulman, D. A., Gaughan, L. C., Unai, T., Casida, J. E., J. Agr. Food Chem. (1976) 24, 270.
5. Gaughan, L. C., Unai, T., Casida, J. E., J. Agr. Food Chem. (1977) in press.
6. Unai, T., Casida, J. E., ACS Symp. Ser. (1977) this volume.

18

Synthesis of Permethrin Metabolites and Related Compounds

TADAAKI UNAI and JOHN E. CASIDA

Pesticide Chemistry and Toxicology Laboratory, Department of Entomological Sciences, University of California, Berkeley, Calif. 94720

Considerable progress has been made in understanding the metabolism of the trans- and cis-isomers of 3-phenoxybenzyl 3-(2,2-dichlorovinyl)-2,2-dimethylcyclopropanecarboxylate (permethrin) in rats (1-3), in cows (3), in insects (4), in bean and cotton plants (3) and in microsomal mixed-function oxidase systems from mammalian liver (5) and insects (4), in part because of the availability of authentic standards from synthesis for use in cochromatographic comparisons with the metabolites. This report outlines synthesis routes used to prepare these monohydroxy- and dihydroxy-derivatives of trans- and cis-permethrin, their hydrolysis products, and certain further oxidized or conjugated derivatives of the hydrolysis products.

3-(2,2-Dichlorovinyl)-2-hydroxymethyl-2-methylcyclopropanecarboxylic Acids

There are 4 possible isomeric acids with hydroxylation at one of the gem-dimethyl positions (Figure 1; the 1R isomers are shown).

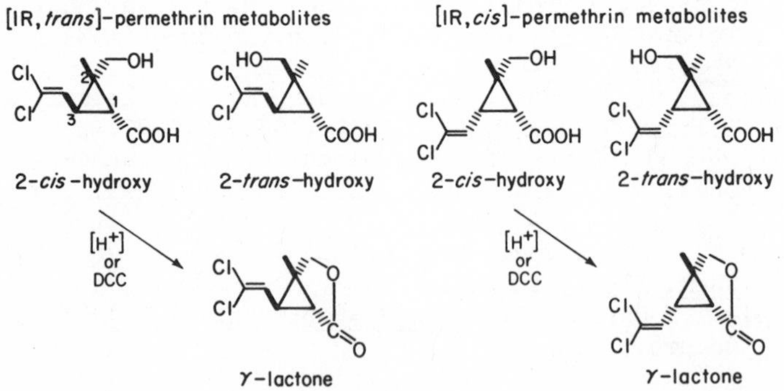

[1R,*trans*]-permethrin metabolites — [1R,*cis*]-permethrin metabolites

2-*cis*-hydroxy — 2-*trans*-hydroxy — 2-*cis*-hydroxy — 2-*trans*-hydroxy

[H+] or DCC → γ-lactone

Figure 1

The 2-cis-hydroxymethyl acids undergo partial conversion to the corresponding γ-lactones under strong acidic conditions and complete conversion on treatment with N,N'-dicyclohexylcarbodiimide (DCC). These 6 compounds are easily isolated as crystalline materials by subjecting the appropriate mixtures to tlc on silica gel chromatoplates using 2 developments with benzene (saturated with formic acid)-ether (10:3) (referred to as the BFE solvent system).

The 4 isomeric hydroxy acids and 2 isomeric γ-lactones were prepared from 1,1-dichloro-4-methyl-1,3-pentadiene by SeO_2 oxidation in glacial acetic acid, which yields the desired dichlorodiene acetates in a ratio of 83 for the trans- and 17 for the cis-compound (Figure 2) plus other products including the dichlorodiene trans-aldehyde.

Figure 2

The mixture of acetoxy compounds was reacted directly with ethyl diazoacetate and Cu powder at 120-130°C, followed by tlc separation [CCl_4-ether (93:7)] of the isomeric diesters. Finally, hydrolysis (NaOH in MeOH), acidification (HCl), tlc (BFE) and lactonization (DCC) yielded the 6 desired [1RS]-compounds. Each of the hydroxy acids and lactones is found as a metabolite of the appropriate permethrin isomer.

Hydroxy Derivatives of 3-Phenoxybenzyl Alcohol and 3-Phenoxybenzoic Acid

Twelve of the 14 possible isomeric monohydroxy derivatives of 3-phenoxybenzyl alcohol and 3-phenoxybenzoic acid were prepared, the remaining 2-hydroxy compounds not being likely metabolites. The 6 benzoic acid derivatives were prepared previously (6), usually by different synthesis routes, while most if not all of the 6 benzyl alcohol derivatives are new compounds. The general procedure involved formation of the diphenyl ether linkage with suitable methoxy intermediates by the Ullmann reaction [Cu_2Cl_2, dimethylformamide (DMF)], demethylation with BBr_3 (CH_2Cl_2, -10°C, 2 min) or $AlCl_3$ (thiophene-free benzene, reflux, 2 hr), and appropriate oxidation or reduction steps to obtain the desired acids and alcohols. The acids are separated by tlc on 2 developments with the BFE solvent system and the alcohols are resolved with benzene-ethyl acetate (6:1) or benzene-ethyl acetate-methanol (15:5:1).

To synthesize the 2'-,3'- and 4'-hydroxy compounds, methyl 3-bromobenzoate was reacted with the sodium salts of 2-, 3- and 4-methoxyphenols (7) (from the phenols and NaH in DMF).

Demethylation and reduction [sodium bis(2-methoxyethoxy)aluminum hydride in benzene (Vitride®)] or hydrolysis (NaOH in MeOH) of the carbomethoxy compounds gave the desired hydroxy alcohols and acids, respectively (Figure 3).

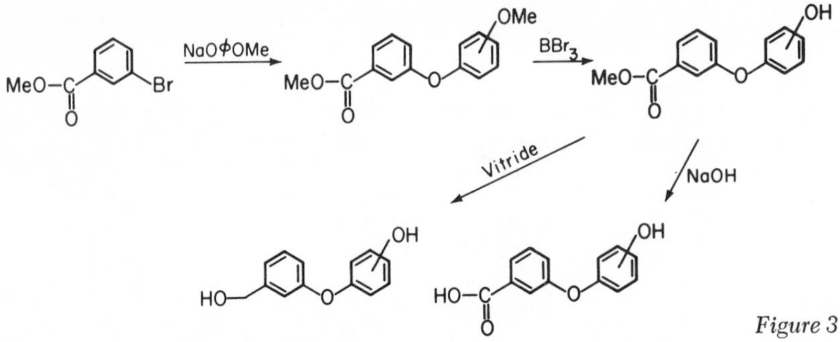

Figure 3

To prepare the 4-hydroxy derivatives, the diphenyl ether aldehyde from an Ullmann reaction of the sodium salt of isovanillin and bromobenzene was oxidized ($KMnO_4$, acetone), demethylated, and the resulting hydroxy acid was converted to the methyl ester (CH_2N_2, ether, 0°C, 5 min) for reduction [$LiAlH_4$ in tetrahydrofuran (THF)] to the desired alcohol (Figure 4).

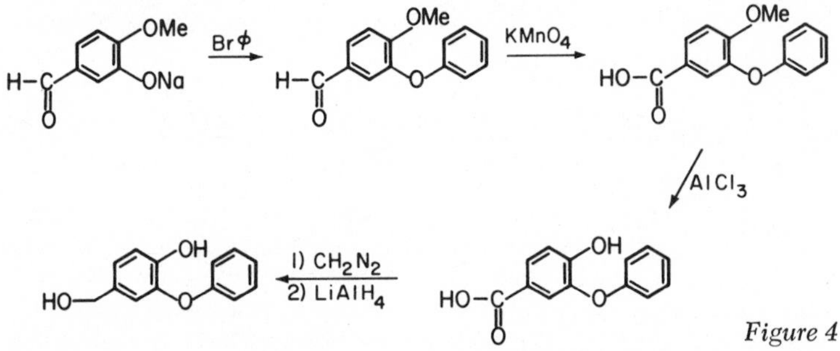

Figure 4

For synthesis of the 5-hydroxy compounds, the sodium salt of 3-hydroxy-5-methoxytoluene was prepared by methylation of 3,5-dihydroxytoluene with dimethyl sulfate, then treatment with NaH. An Ullmann reaction of this salt with bromobenzene gave 5-methoxy-3-phenoxytoluene (6), which was subjected to oxidation [$KMnO_4$, H_2O-pyridine (1:2)] to the methoxy acid and further reactions (Figure 5) as above.

The 6-hydroxy derivatives were prepared by methylation of 5-bromosalicylaldehyde with dimethyl sulfate then reaction of the methoxy bromide with sodium phenolate and oxidation ($KMnO_4$, acetone) of the aldehyde and further treatments as before (Figure 6).

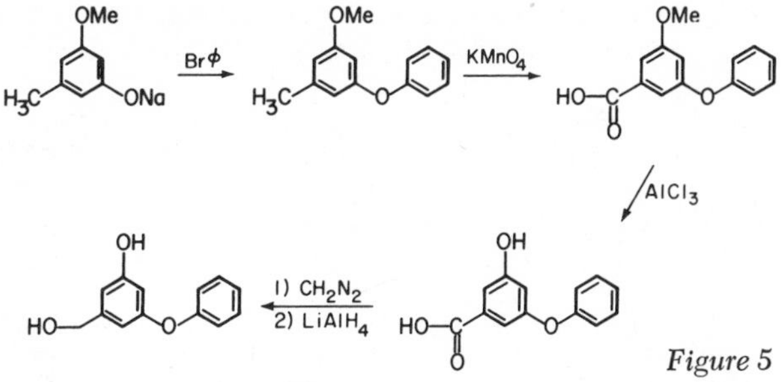

Figure 5

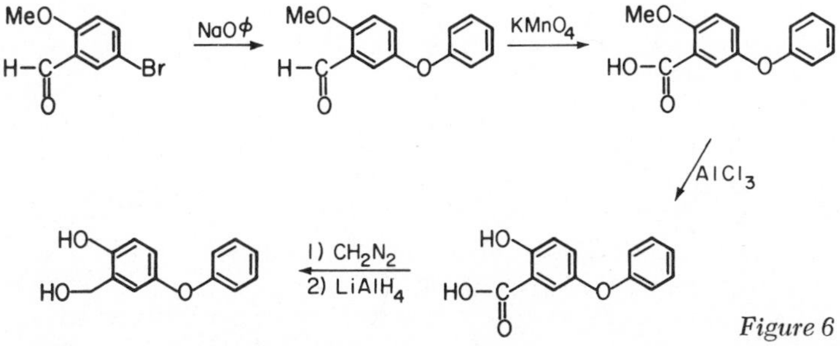

Figure 6

The 2'-, 4'- and 6-hydroxy-3-phenoxybenzoic acids and the corresponding 4'- and 6-hydroxy benzyl alcohols in free or conjugated form appear as metabolites of trans- and/or cis-permethrin in various biological systems.

Mono- and Dihydroxypermethrin

The trans- and cis-isomers of permethrin with monohydroxylation in the acid moiety were synthesized by heating the 4 isomeric hydroxy acids (Figure 1) with 3-phenoxybenzyl bromide and Et_3N in DMF solution in ampoules with N_2 gas at 80-90°C for 2-3 hr (Figure 7). Each product was purified by preparative tlc with benzene-ethyl acetate (6:1) and benzene-ethyl acetate-methanol (15:5:1). Isomerization was not observed under these conditions of reaction or purification but a small amount of lactonization occurred with the 2-cis-hydroxymethyl compounds.

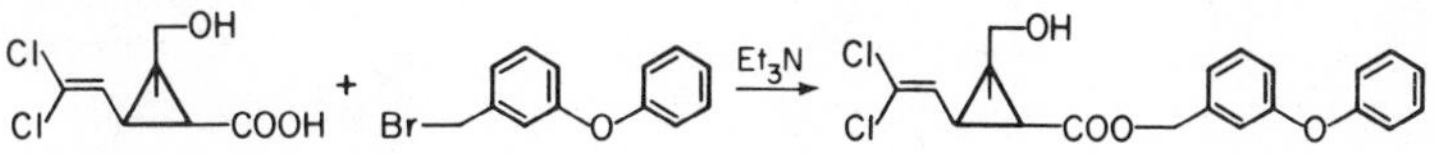

Figure 7

For synthesis of the cis-esters with monohydroxylation at the 2'- and 4'-positions of the phenoxy ring and the corresponding trans-ester with a 4'-hydroxy substituent, the appropriate methoxy esters prepared by the acid chloride method were treated with BBr_3 (CH_2Cl_2, -10°C, 2 min) to yield the desired monohydroxy esters as minor products and the dichlorovinyl acid and the corresponding hydroxy derivatives of 3-phenoxybenzyl bromide as major products, using preparative tlc with benzene (for the 2'-hydroxy ester) or with benzene-ethyl acetate (6:1) (for the 4'-hydroxy esters) for the isolations. Reesterification of the 2 isomeric hydroxy derivatives of 3-phenoxybenzyl bromide with either the trans- or cis-dichlorovinyl acid yielded the desired ester (including the trans-ester with a 2'-hydroxy substituent) in each case (Figure 8).

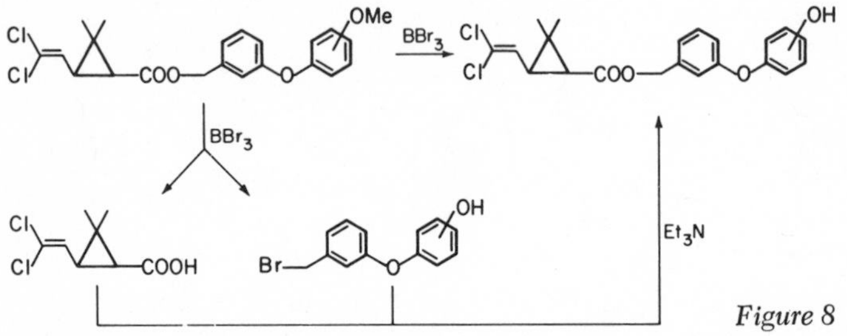

Figure 8

Esters hydroxylated at both the 2-trans-methyl of the acid moiety and the 4'-position of the alcohol moiety (Figure 9) were prepared by esterifying the 2 isomeric acids with hydroxylation at the trans-methyl to the carboxyl (Figure 1) with 4'-hydroxy-3-phenoxybenzyl bromide (Figure 9).

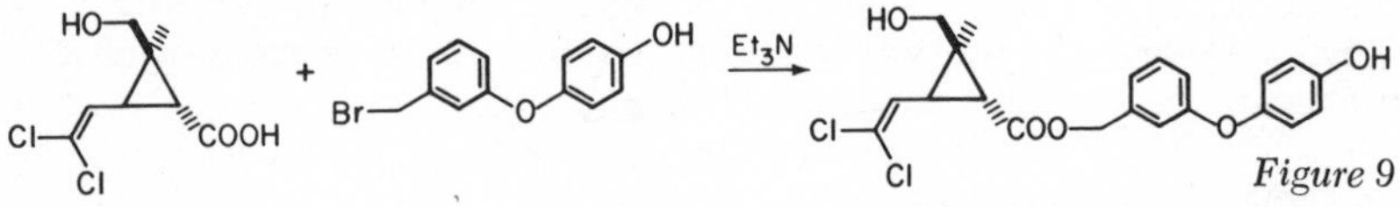

Figure 9

Each of the mono- and dihydroxy derivatives of permethrin described above (with the exception of the 2'-hydroxy-trans-ester) is a metabolite of trans- or cis-permethrin in one or more of the biological systems examined.

Amino Acid and Sulfate Conjugates

Twelve amino acid conjugates were prepared as their methyl esters from the acid chlorides of the trans- and cis-dichlorovinyl acids or of 3-phenoxybenzoic acid and the methyl esters of the L-amino acids with pyridine in THF-benzene solution (Figure 10). The trans- and cis-dichlorovinyl acid metabolites from permethrin are conjugated with glycine, serine and glutamic acid in insects and the trans- but not the cis-acid is detected as a glutamate conjugate with cows. Phenoxybenzoic acid is conjugated with glycine in rats, cows and insects and with glutamic acid in cows and insects.

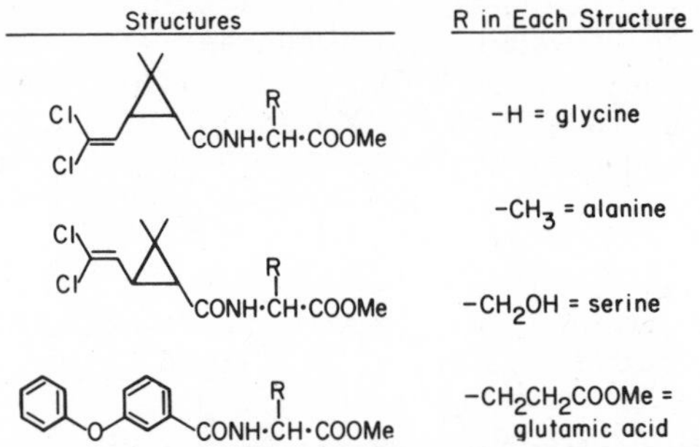

Figure 10

The sulfate conjugate of 4'-hydroxy-3-phenoxybenzoic acid, found as a major permethrin metabolite in rats, was prepared by sulfation of the acid with $ClSO_3H$ in pyridine solution (Figure 11). The product obtained in poor yield was purified by preparative tlc with n-butanol-acetic acid-H_2O (6:1:1). The starting material was obtained on hydrolysis of this sulfate with sulfatase or 3N HCl.

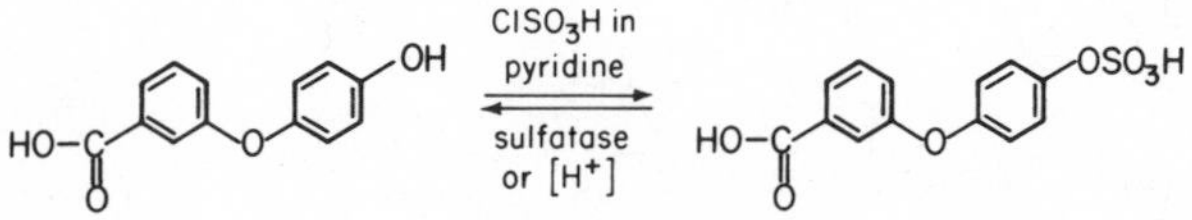

Figure 11

Abstract

Mono- and dihydroxy derivatives of [1RS]-trans- and [1RS]-cis-permethrin, their ester hydrolysis products, and conjugates of the acid moieties and of 3-phenoxybenzoic acid and 4'-hydroxy-3-phenoxybenzoic acid were prepared for verification and stereochemical assignments of the free and conjugated [^{14}C]-metabolites of the [^{14}C]permethrin isomers. At least 2 different solvent systems were used in each case for cochromatographic identification, with and without derivatization of the compounds. Twenty-nine of the products synthesized are identified as permethrin metabolites in free or conjugated form. These compounds were important in assigning structures for the permethrin metabolites formed in various organisms and enzymatic systems. They should also be useful standards in studies on metabolism of related pyrethroids.

Acknowledgments

The authors thank Loretta Gaughan, Roy Holmstead, Toshio Shono, David Soderlund and Kenzo Ueda for valuable suggestions and assistance. This study was supported in part by grants from: National Institutes of Health (2 P01 ES00049); Agricultural Chemical Div., FMC Corp., Middleport, N.Y.; Agricultural

Chemicals Div., ICI United States Inc., Goldsboro, N. C.; Sumitomo Chemical Co., Osaka, Japan; Roussel-Uclaf-Procida, Paris, France; Mitchell Cotts & Co. Ltd., London, England; Wellcome Foundation Ltd., London, England; National Research Development Corp., London, England.

Literature Cited

1. Elliott, M., Janes, N. F., Pulman, D. A., Gaughan, L. C., Unai, T., Casida, J. E., J. Agr. Food Chem. (1976) 24, 270.
2. Gaughan, L. C., Unai, T., Casida, J. E., J. Agr. Food Chem. (1977) in press.
3. Gaughan, L. C., Unai, T., Casida, J. E., ACS Symp. Ser. (1977) this volume.
4. Shono, T., Unai, T., Casida, J. E., unpublished results.
5. Soderlund, D. M., Casida, J. E., ACS Symp. Ser. (1977) this volume.
6. Ungnade, H. E., Rubin, L., J. Org. Chem. (1951) 16, 1311.
7. Miyamoto, J., Suzuki, T., Nakae, C., Pestic. Biochem. Physiol. (1974) 4, 438.

19

Synthetic Pyrethroids: Residue Methodology and Applications[1]

D. A. GEORGE, J. E. HALFHILL, and L. M. McDONOUGH

Yakima Agricultural Research Laboratory, Agricultural Research Service, U.S. Department of Agriculture, Yakima, Wash. 98902

The shortcomings of DDT [1,1,1-trichloro-2,2-bis=(p-chlorophenyl)ethane] and other chlorinated hydrocarbons has stimulated research into the development of other broad spectrum insecticides that could provide high insecticidal activity combined with low mammalian toxicity and moderate persistence. Compounds related to natural pyrethrum (synthetic pyrethroids) have been developed in recent years that show promise of meeting these requirements. Natural pyrethrum (Fig. 1) is an ester with synthetic variations of both the acid and alcohol portion of the molecule providing promising insecticides.

Field studies indicate that Bioethanomethrin® [(5-benzyl-3-furyl)methyl trans-(+)-3-(cyclopentylidenemethyl)-2,2-dimethylcyclopropanecarboxylate] and FMC 33297 [m-phenoxybenzyl cis,trans-(±)-3-(2,2-dichlorovinyl)-2,2-dimethylcyclopropane=carboxylate] (Fig. 2), showed promise of controlling insects of certain vegetable crops. Consequently, we investigated methods of detecting their residues. We report here a direct method utilizing gas chromatography and electron capture detection for FMC 33297. In addition, we developed a method based on derivatives of the alcohol and acid moieties of the saponified pyrethroid molecule for both pyrethroids (1). This latter method should work equally well with other synthetic pyrethroids (Fig. 3).

The sample is saponified, the alcohol moiety is extracted with methylene chloride, the water portion is then acidified, and the acid moiety is extracted with methylene chloride. Trichloroacetyl chloride with pyridine is used to form a trichloroacetate ester from the alcohol and trichloroethanol

[1] This paper reports the results of research only. Mention of a pesticide in this paper does not constitute a recommendation for use by the U.S. Department of Agriculture nor does it imply registration under FIFRA as amended.

CHRYSANTHEMATES **PYRETHRATES**

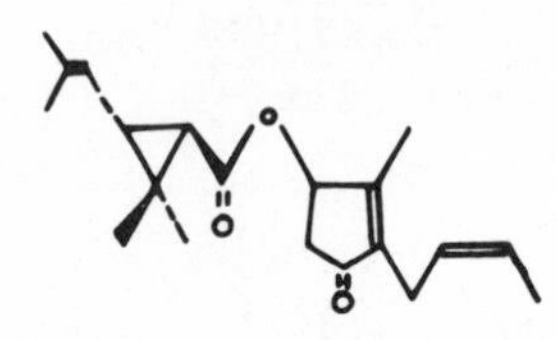

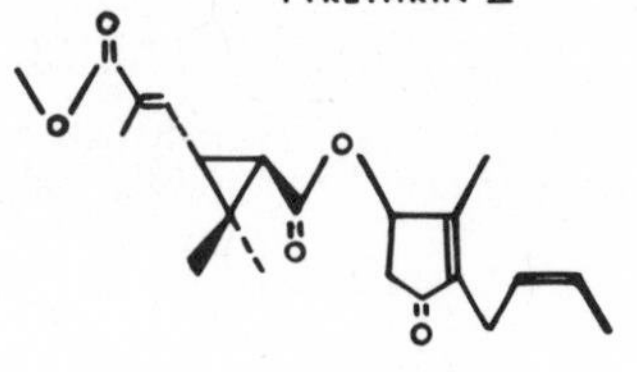

PYRETHRIN I PYRETHRIN II

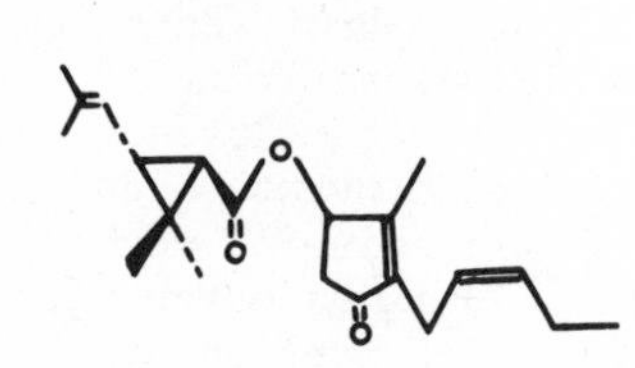

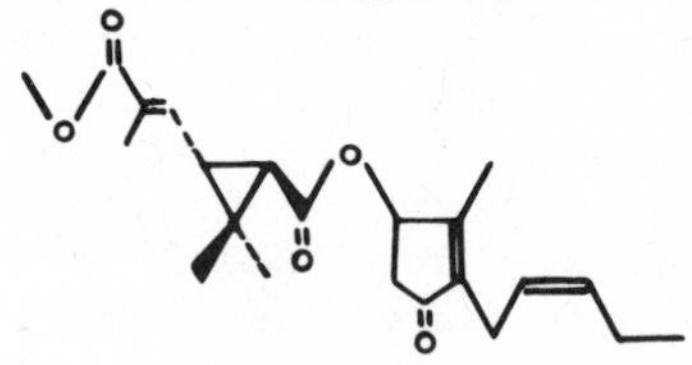

CINERIN I CINERIN II

JASMOLIN I JASMOLIN II

Figure 1. Natural pyrethrums: chrysanthemates and pyrethrates

Cl

Cl

O

O

Figure 2. Synthetic pyrethroids Bioethanometrin and FMC 33297

with pyridine and dicyclohexylcarbodiimide is used to form an ester with the acid. This use of a trichloro compound in the derivatives increases the sensitivity to the nanogram range when the derivatives are determined by a gas chromatograph equipped with an electron capture detector.

The infrared spectra support the expected structures for the derivatized products (Fig. 4). The carbonyl stretching vibrational frequency is at 5.68 microns for the trichloroacetate of both pyrethroids, the aromatic ether stretching frequency is at 8 microns, and the carbon-chloride bands are near 14 and 17 microns.

The esters from the acid moiety for both pyrethroids show the carbonyl stretching frequency at 5.78 microns, a strong C-O-C triplet at 8.7, 8.9, and 9.1 microns (Fig. 5). Again, the carbon-chloride stretching frequencies are at 14 and 18 microns.

The high resolution mass spectra confirm the expected structures of the derivatives formed (Fig. 6). The molecular ion for the trichloroacetate ester of the Bioethanomethrin derivative was found to be m/e 331.9789 (calculated m/e 331.9773). The molecular ion of the trichloroethanol ester was found to be m/e 324.0445 (calculated m/e 324.0449). The base peak of the trichloroacetate ester was found at m/e 171.0804, corresponding to the loss of the trichloroacetate ion from the molecule. The base peak of the trichloroethanol ester was found at m/e 149.1323, corresponding to the loss of $C_3H_2O_2Cl_3$ ($CO_2CH_2CCl_3$).

The molecular ion for the trichloroacetate ester derivative of FMC 33297 was m/e 346.9764 (calculated m/e 345.9765). The molecular ion of the trichloroethanol ester derivative was m/e 337.9185 (calculated m/e 337.9200) (Fig. 7). The base peak of the trichloroacetate ester, corresponding to the loss of the trichloroacetate ion from the molecule, was found at m/e 183.0808. The base peak of the trichloroethanol ester, corresponding to the loss of $CO_2CH_2CCl_3$, was found at m/e 163.0058.

In addition, a method was developed for determining residues of FMC 33297 in which the complete molecule was used. The 2 chlorine atoms in the molecule make the compound sensitive to the electron capture detector; the result is a standard curve with a sensitivity of 5 to 50 nanograms. However, we obtained a 10-fold greater sensitivity with the derivatization method.

Residue Studies

An emulsifiable concentrate of Bioethanomethrin was applied to lentils as a spray 2 times at the rate of 700 g AI/hectare (10 oz/acre) 14 and 21 days before harvest. Foliage residues were collected after the first spray at 0 hr, 66 hr, and 168 hr, and after the 2nd treatment at 0 hr and 72 hr. The samples were selected randomly throughout the plot, placed

$$R^1\text{-}\overset{O}{\overset{\|}{C}}\text{-}O\text{-}CH_2\text{-}R^2 \xrightarrow[2)\ H_3O^+]{1)\ OH^{\ominus},\ MeOH} \begin{cases} R^2\text{-}CH_2OH \xrightarrow{Cl_3C\text{-}COCl} R^2CH_2\text{-}O\text{-}\overset{O}{\overset{\|}{C}}\text{-}CCl_3 \\ R^1\text{-}COOH \xrightarrow[\text{pyridine - DCC}]{Cl_3\ C\text{-}CH_2OH} R^1\text{-}\overset{O}{\overset{\|}{C}}\text{-}O\text{-}CH_2\text{-}Cl_3 \end{cases}$$

Figure 3. Derivative step (DCC is dicyclohexylcarbodiomide)

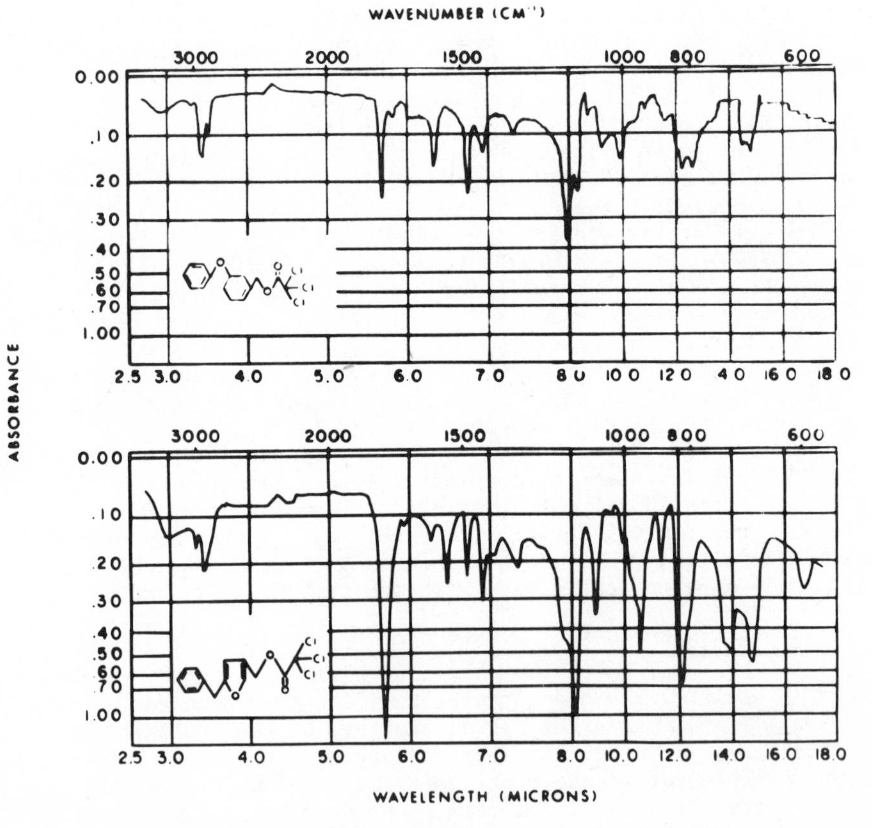

Figure 4. Infrared spectra of trichloroacetate ester derivatives

in plastic bags, and frozen at the laboratory. The frozen samples were put through a chopper and mixed thoroughly; 50 g subsamples were removed for analysis, blended 2 minutes with 200 ml methylene chloride, and filtered through anhydrous sodium sulfate. The filtrate was refrigerated until analyzed. The foliage extract was cleaned up by evaporating the methylene chloride, redissolving the residue in hexane, and then subjecting it to liquid chromatography with an aluminum oxide column (20 g of Baker's Analyzed 0536). After 50 ml of hexane was passed through the column, Bioethanomethrin was eluted with 1.5% acetonitrile in hexane.

The lentils were also analyzed. Extraction was the same as for the foliage. Then the extract was evaporated to dryness, and the residue was partitioned between hexane and acetonitrile. The acetonitrile solution was evaporated and the residue was redissolved in hexane and chromatographed in the same manner as the foliage sample.

The residues in foliage determined from the 2 derivatives compare favorably. The rate of loss during the 7-day sampling is shown in Table I. Immediately after spraying, the residues were 0.4 to 0.6 ppm and declined to zero by the end of the 168-hr sampling time. Control or check samples showed no residue by either method. Recoveries (in the range of 0.1 to 1.0 ppm) averaged 84.2% for the trichloroacetate ester derivative and 80.3% for the trichloroethanol ester derivative. No residues were found in the harvest lentils.

Table I. Residues of Bioethanomethrin on lentil foliage at various intervals after an application of 700 g AI/hectare (10 oz/acre).

Interval between treatment and sampling (hr)	Residues found (ppm)[1/]	
	Trichloroacetate ester derivative	Trichloroethanol ester derivative
	Treatment 1	
0	0.350	0.447
66	.035	.066
168	.000	.000
	Treatment 2	
0	.647	.689
72	.000	.000

1/ Results were corrected to 100% based on recoveries.

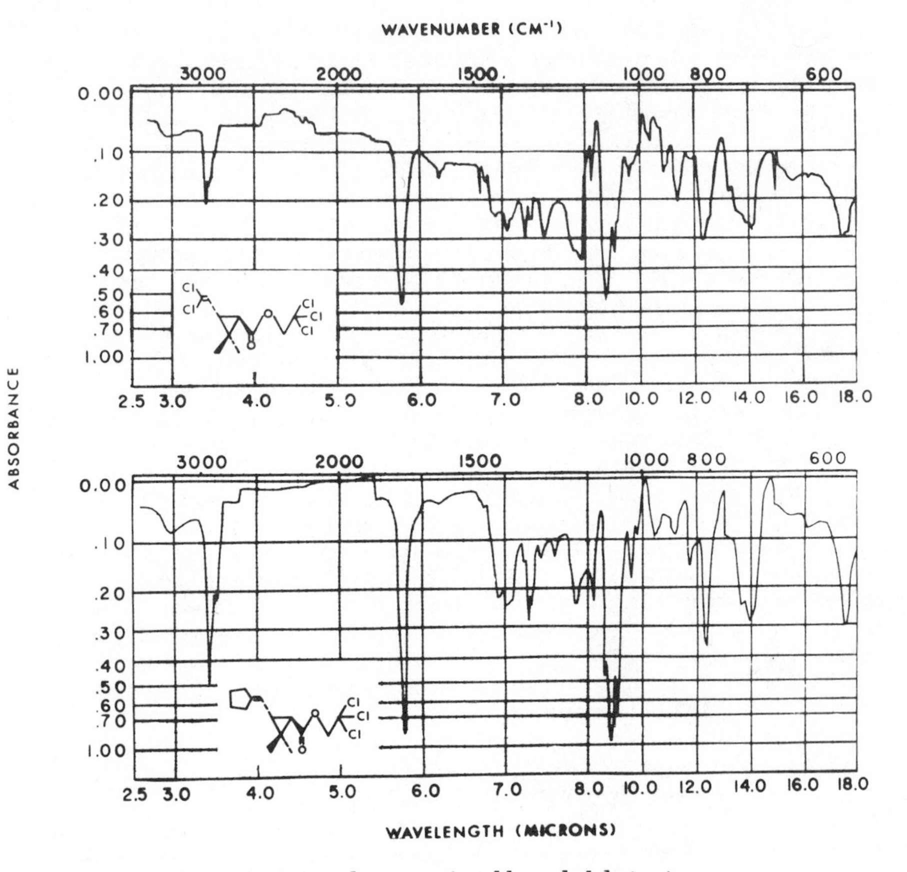

Figure 5. Infrared spectra of trichloroethyl derivatives

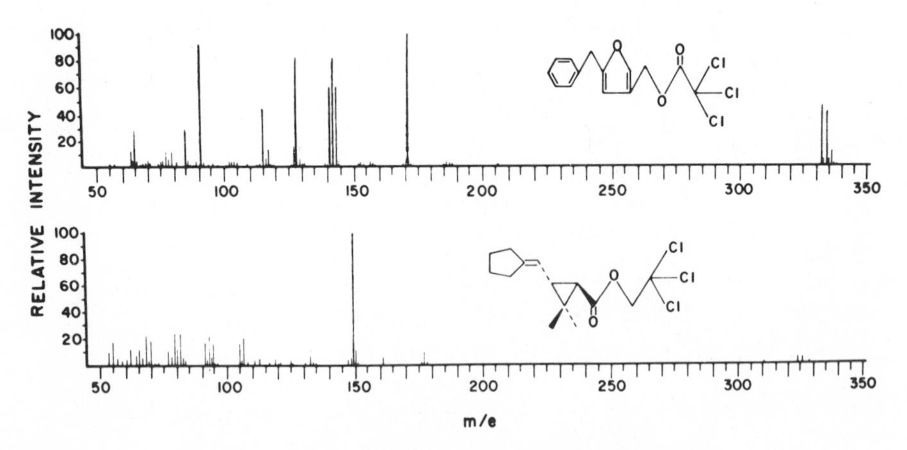

Figure 6. Mass spectra of Bioethanomethrin derivatives

We also applied an emulsifiable concentrate of FMC 33297 to cabbage once a week for 4 weeks at a rate of 112 g AI/hectare (0.1 lb/acre). Sampling and extraction were the same as for the lentil and foliage samples treated with Bioethanomethrin. Cleanup of the extract consisted of liquid chromatography of 40 ml of a methylene chloride solution (10 g crop sample) through a 7.5 g silica gel column, prewashed with 30 ml methylene chloride. FMC 33297 residues were eluted with 20 ml more of methylene chloride. After evaporation of the eluate, the residue was dissolved in pentane and liquid chromatographed through aluminum oxide (20 g). After 50 ml pentane had been passed through the column, the FMC 33297 residues were eluted with 1.5% ether in pentane. The eluate was evaporated, and 1 ml hexane was added. After the residue was determined by electron capture gas chromatography, the sample was also analyzed by the derivatization method.

Residues determined by the 3 techniques are shown in Table II. By Duncan's multiple range test, there was no significant difference between the 3 methods at the 95% probability level. Immediately after spraying, the residue in the outer leaves was 1.4 ppm and declined approximately 30% in the first 24-hr period and 70% within 168 hr. No residues were found in the cabbage heads 168 hr after treatment. Check samples showed no residues for the 3 methods. Recoveries (in the range of 0.04 to 2.0 ppm) averaged 102.3% for underivatized FMC 33297, 98.9% for the trichloroacetate ester derivative and 92.5% for the trichloroethanol ester derivative.

We applied FMC 33297 to green peas at the rate of 112 g AI/hectare (0.1 lb/acre). Samples consisting of vine and pod were taken at 0 hr, 24 hr, 3 days and 7 days. The sampling, extraction, and cleanup procedures are the same as those used for cabbage. The results are shown in Table III. Initial residues were about 0.9 ppm and declined approximately 30% during the first 24-hr period, 50% by 72 hr, and 80% after 168 hr. Check samples showed no residues for the 3 methods. Recoveries (in the range of 0.1 to 1.0 ppm) averaged 101.8% for the underivatized FMC 33297, 84.6% for the trichloroacetate ester derivative, and 89.7% for the trichloroethanol ester derivative.

Comparison of the results by the derivatization methods thus indicates that either procedure can be used to determine residues; the other method is then available for corroboration. Also, when a faster method is available, as for FMC 33297, the derivatization technique can be used for corroboration. The data we obtained indicate that FMC 33297 residues are more persistent than Bioethanomethrin residues.

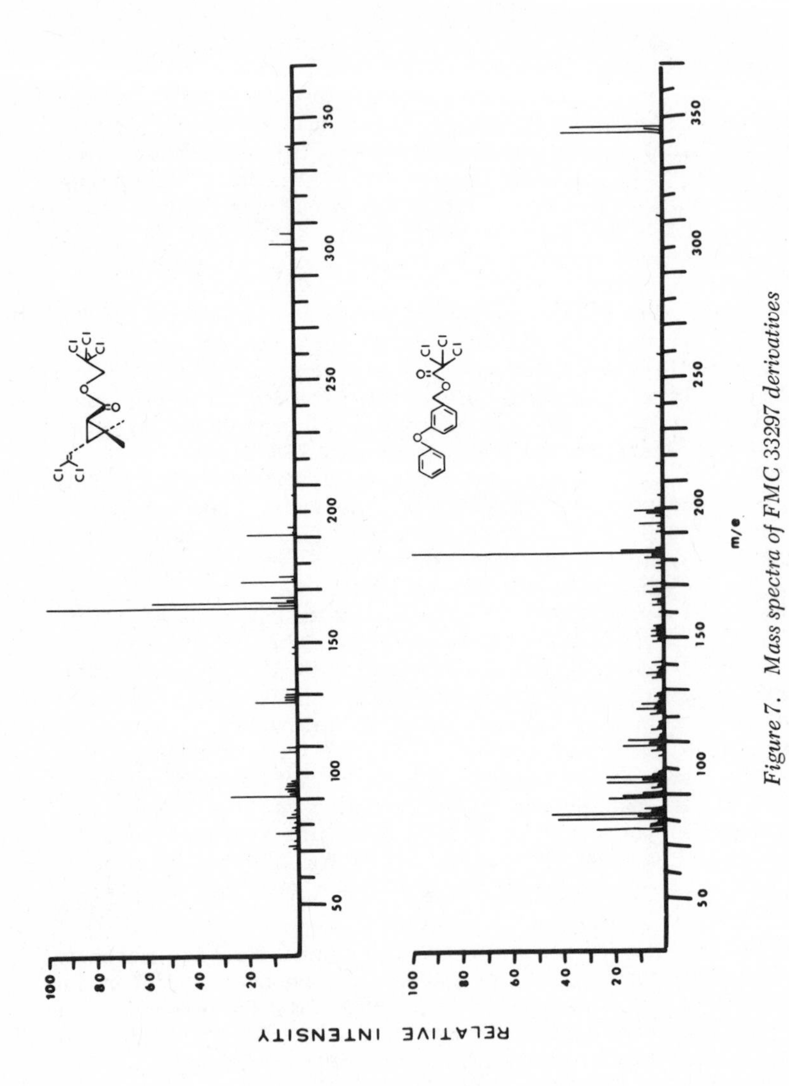

Figure 7. Mass spectra of FMC 33297 derivatives

Table II. FMC 33297 residues on outer cabbage leaves after 4 spray applications once a week of 112 g AI/hectare (0.1 lb/acre).

Interval between treatment and sampling (hr)	No. of treatments	Residue found (ppm)[1/] Underivatized FMC 33297	Trichloro-acetate ester derivative	Trichloro-ethanol ester derivative
0	1	1.20	1.23	1.43
	2	2.29	2.61	2.81
	3	.95	1.17	.76
	4	.92	.63	.52
	Avg.	1.34	1.41	1.38
24	1	1.25	1.01	1.16
	2	.86	.99	1.25
	3	.77	.23	.55
	4	1.38	.77	.92
	Avg.	1.07	.75	.97
168	2	0.10	0.06	0.06
	3	.32	.42	.32
	4	.72	.80	.73
	Avg.	.38	.43	.37

1/ Results were corrected to 100% based on recoveries found

Table III. FMC 33297 residues on green peas after a spray application of 112 g AI/hectare (0.1 lb/acre).

Interval between treatment and sampling (hr)	Residues found (ppm)[1/]		
	Underivatized FMC 33297	Trichloro-acetate ester derivative	Trichloro-ethanol ester derivative
0	0.98	0.60	0.96
24	.69	.55	.60
72	.59	.39	.28
168	.18	.13	.15

1/ Results were corrected to 100% based on recoveries found.

Abstract

Analytical methodology was developed for determining residues of 2 synthetic pyrethroids, Bioethanomethrin® [(5-benzyl-3-furyl)methyl trans-(+)-3-(cyclopentylidenemethyl)-2,2-dimethylcyclopropanecarboxylate] and FMC 33297 [m-phenoxybenzyl cis,trans-(±)-3-(2,2-dichlorovinyl)-2,2-dimethylcyclopropane=carboxylate]. After saponification of a pyrethroid, trichloroacetyl chloride is used to form an ester from the alcohol moiety, and trichloroethanol is used to form an ester from the acid moiety. The derivatives were determined by electron capture gas chromatography. Infrared and mass spectral data support the expected structures of the derivatives formed. The use of the 2 derivatives enhances the reliability of the results. In addition, underivatized FMC 33297 was determined by electron capture gas chromatography. Equivalent residue data were obtained by the 3 analytical techniques. With these methods, residue data were determined for 2 pyrethroids (Bioethanomethrin and FMC 33297) on lentil foliage, lentils, green peas, and cabbage. The data indicate that residues declined 70 to 100% within 7 days of treatment.

Literature Cited

1. George, D. A., and McDonough, L. M., J. Assoc. Off. Anal. Chem. (1975) 58, 781-4.

20

Gas Chromatographic Determination of Residues of the Synthetic Pyrethroid FMC 33297

R. A. SIMONAITIS and R. S. CAIL

Stored-Product Insects Research and Development Laboratory, Agricultural Research Service, U.S. Department of Agriculture, Savannah, Ga. 31403

FMC 33297 (m-phenoxybenzyl cis,trans-(±)-3-(2,2-dichloro=vinyl)-2,2-dimethylcyclopropanecarboxylate), also known as NRDC 143, PP 557, and permethrin, is a photostable synthetic pyrethroid with low mammalian toxicity which has been shown by Bry and Lang (1)to have promise as a protectant for woolen fabrics against insect damage. Davis et al. (2) have found FMC 33297 to be promising against insect pests of growing cotton. Gillenwater and his coworkers at the Stored-Product Insects Research and Development Laboratory in Savannah, Georgia (private communication), have found FMC 33297 to be an effective protectant for stored grain commodities. The compound (structural formula shown in Figure 1) was synthesized by Elliott et al. (3) in their search for synthetic pyrethroids having low mammalian toxicity and more stability than the natural pyrethroids that were nevertheless equally active against insects. Berkovitch (4) reported that FMC 33297 was 1.4 times as potent as resmethrin ((5-benzyl-3-furyl)methyl cis,trans-(±)-2,2-dimethyl-3-(2-methyl=propenyl)cyclopropanecarboxylate) to house flies, *Musca domestica* L., and 3.2 times as effective to mustard beetles, *Phaedon cochicariae* Fab., and that it had comparable mammalian toxicity. Elliott et al. (3), Berkovitch (4), and Bry et al. (5) reported that FMC 33297 was many times more stable when exposed to light than previously synthesized pyrethroids or natural pyrethrins.

Preliminary experiments showed that FMC 33297 was effective against various insect pests of stored grains. Consequently, a

---

[1]This paper reports the results of research only. Mention of a pesticide in this paper does not constitute a recommendation for use by the U.S. Department of Agriculture nor does it imply registration under FIFRA as amended. Mention of a commercial or proprietary product in this paper does not constitute an endorsement of this product by the U. S. Department of Agriculture.

method was needed to determine residues of FMC 33297 on unmilled grains such as wheat and corn and on milled products such as cornmeal and flour so we could study application rates and duration of protection afforded the commodity. This paper describes a highly sensitive gas-liquid chromatographic (GLC) procedure for determining residues of the *cis* and the *trans* isomers of FMC 33297 on milled and unmilled corn and wheat.

Method

Apparatus and Reagents.

(a) Gas chromatographic equipment and conditions.--Hewlett Packard model 5700A instrument with automatic injector, digital integrator, 1 mv recorder (Hewlett Packard Co., Avondale, PA 19311); equipped with flame ionization detector and 122 cm x 4 mm id glass column packed with 5% liquid phase OV-225 w/w on 80-100 mesh Gas-Chrom Q (Applied Sciences Laboratories, Inc., State College, PA 16801). Condition a newly packed column 24 hr at 300°C with nitrogen purging. Operating conditions: temperature (°C) column oven 250, detector 300, injection port 300; gas flows (ml/min) nitrogen (carrier) 28, hydrogen 30, air 240; electrometer setting, 31 x 10^{-10} A full scale at attenuation 16, Range 1. Under these conditions the retention times for the *cis* and *trans* isomers of FMC 33297 were 9.5 min and 10.0 min, respectively.

(b) Chromatographic column.--For alumina column cleanup: Chromaflex, plain, 23 cm x 13 mm with 8 cm x 2 mm bore capillary tip, 50 ml reservoir (Kontes Glass Co., Vineland, NJ 08360).

(c) Alumina, Acid.--Brockman Activity I 80-200 mesh; use as received (Fisher Scientific Co., Pittsburgh, PA 15235).

(d) Analytical pesticide standards.--FMC 33297, analytical grade (FMC Corporation, Middleport, NY 14105). Primary standard (1000 µg/ml).--Weigh 0.1±0.001 g into 100 ml volumetric flask and dilute to volume with hexane. Secondary standards.--Dilute aliquots of primary standard with hexane to obtain secondary standards. Concentration of secondary standards should bracket the expected concentration of samples (20-25 µg/ml).

(e) Reagents and solvents.--Acetonitrile, ethyl acetate, hexane, and pentane (all pesticide grade); and anhydrous sodium sulfate (Fisher Scientific Co.).

(f) Eluant mixture.--For alumina liquid chromatographic cleanup.--Dry all solvents and solvent mixture before adding to column by shaking with anhydrous sodium sulfate (ca 25 g/L). Three percent ethyl acetate in pentane (v/v): Add 30 ml ethyl acetate to ca 900 ml pentane in 1 L volumetric flask, mix, let mixture reach room temperature, and adjust to final volume of 1 L with pentane.

Preparation of Sample and Extraction. Grind grain to fine consistency in food homogenizer. Transfer 200±0.1 g thoroughly mixed commodity to 1 qt mason jar, add 400 ml pentane, seal, and extract by tumbling for 3 hr at 10 rev/min (tumbler specifications available from U.S. Department of Agriculture, Stored-Product Insects Research and Development Laboratory, Savannah, GA 31403) or equivalent. Filter extract through Whatman 2V filter paper into 1 pt bottle, and stopper. At this point extracts can be stored in freezer at -5°C until ready for further analysis.

Solvent Partition Cleanup. Saturate acetonitrile with pentane by shaking 3:1 mixture of solvents for 1 min. For each sample to be analyzed, set up two 250 ml separatory funnels. Transfer 50 ml aliquot of extract that is equivalent to 25 g commodity to 125 ml Erlenmeyer flask and evaporate pentane to about 10 ml on 60°C water bath under a stream of dry air. With pentane, quantitatively transfer the concentrated sample in flask to separatory funnel that has been calibrated with a 25 ml mark. Dilute volume to the 25 ml mark with pentane. Add 100 ml pentane-saturated acetonitrile, stopper, and shake 1 min. Transfer lower (acetonitrile) phase to second separatory funnel. Reextract the acetonitrile in the second separatory funnel with an additional 25 ml of pentane. Drain acetonitrile phase from the second separatory funnel into 250 ml Erlenmeyer flask. Combine pentane phases in second separatory funnel. Rinse first separatory funnel with 50 ml pentane-saturated acetonitrile. Transfer rinses to second separatory funnel containing pentane phases, stopper, and shake 1 min; then let layers separate. Add lower (acetonitrile) phase to 250 ml Erlenmeyer flask containing the acetonitrile. Discard pentane layer. Place 250 ml Erlenmeyer flask on 100°C water bath and evaporate just to dryness under dry air stream. Dissolve residue in 10 ml pentane, which has been dried by shaking with anhydrous sodium sulfate, and hold at -5°C for liquid chromatographic column cleanup.

Liquid Chromatographic Cleanup. Prepare chromatographic column containing glass wool plug and 11 g alumina. Tap sides of column to produce even packing of adsorbent. Place 250 ml beaker under column and wet column with ca 10 ml pentane which has been dried with anhydrous sodium sulfate. Transfer sample from solvent partition cleanup to chromatographic column and wash sides with ca 15 ml dry pentane. Discard pentane washings. Elute column with 75 ml 3% ethyl acetate in pentane and collect eluate in 125 ml Erlenmeyer flask. Concentrate eluate just to dryness on 60°C water bath under stream of dry air. Dissolve residue in hexane and transfer to a calibrated centrifuge tube or volumetric flask of the appropriate volume to give the desired concentration of 20 to 25 μg/ml. Fill a 2 ml vial 2/3 full with

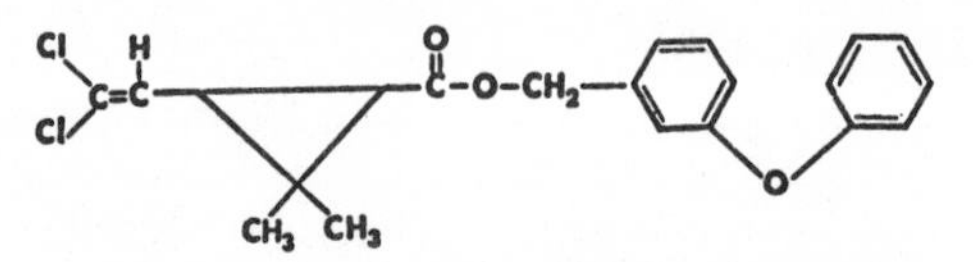

Figure 1. Structural formula of FMC 33297: $C_{21}H_{20}Cl_2O_3$; 391.31 g/mol; LD_{50}; CA 2000 (approximately 20% cis, *80%* trans *isomers)*

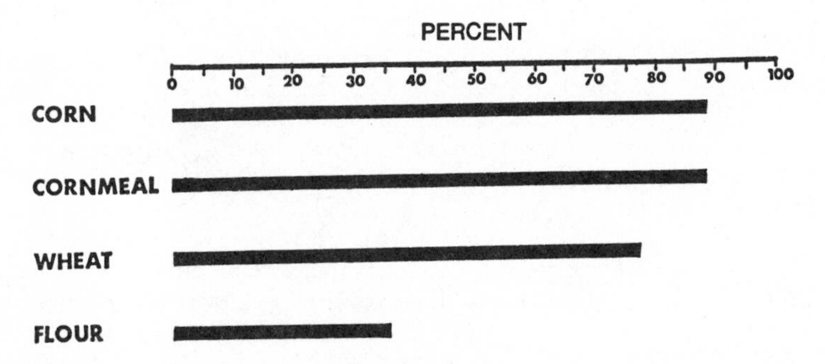

Figure 2. Effectiveness of solvent partition cleanup. Average % nonvolatile material removed by solvent partition from 25 g of commodity extract.

an aliquot of the sample and cap the vial with an aluminum septum cap. Store in a freezer at -5°C until ready for gas chromatographic analysis.

Gas Chromatographic Analysis. Bring all solutions to room temperature. Adjust GLC operating conditions as described under Gas chromatographic equipment and conditions. With automatic sampler, inject 3 μl aliquots of analytical standard solution until integrator counts of two consecutive injections vary <10%. Place vials containing analytical standard before and after vials that contain sample solutions. Inject each twice with automatic sampler. Average integrator counts and calculate percentage FMC 33297 as follows:

$$\% \text{ FMC } 33297 = \frac{W_s}{C_s} \times \frac{C_u}{W_u} \times D \times 100$$

where W_s = weight per cent of standard; W_u = weight percent of unknown; C_s = average integrator counts for standards; C_u = average integrator counts for unknown; D = dilution factor.

Results and Discussion

The cleanup procedure described in this paper utilizes a liquid-liquid partitioning between pentane and acetonitrile followed by a liquid chromatographic cleanup on an acid alumina column. With the extraction procedure described, a 25 g sample of corn, cornmeal, flour, and wheat had nonvolatile residues of 0.77, 1.14, 0.11, and 0.26 g, respectively. Figure 2 shows the quantities of nonvolatile residues removed from extracts of each commodity by solvent partition. No nonvolatile residue was discerned after the liquid chromatographic cleanup.

Figure 3 shows gas chromatograms obtained for an FMC 33297 standard solution before (I) and after (II) cleanup. The retention times for the elution of the *cis* and *trans* isomers of FMC 33297 are indicated by A and B, respectively. The peaks identified by C originated from impurities in the solvents used. A reagent blank shown in Figure 4 was run to determine whether the reagents could be a source for interference. Although peaks were observed, no impurity peak had a retention time which would interfere with the analysis.

The linearity of the detector response was determined by plotting the integrator counts obtained versus the concentration of FMC 33297 injected. The volume injected was kept constant. The response of the flame ionization detector to FMC 33297 was found to be linear over the range examined from 15 to 485 ng per 3 μl injection volume.

Untreated corn, cornmeal, flour, and wheat and samples fortified with 0.2 to 22.0 ppm of FMC 33297 were extracted and

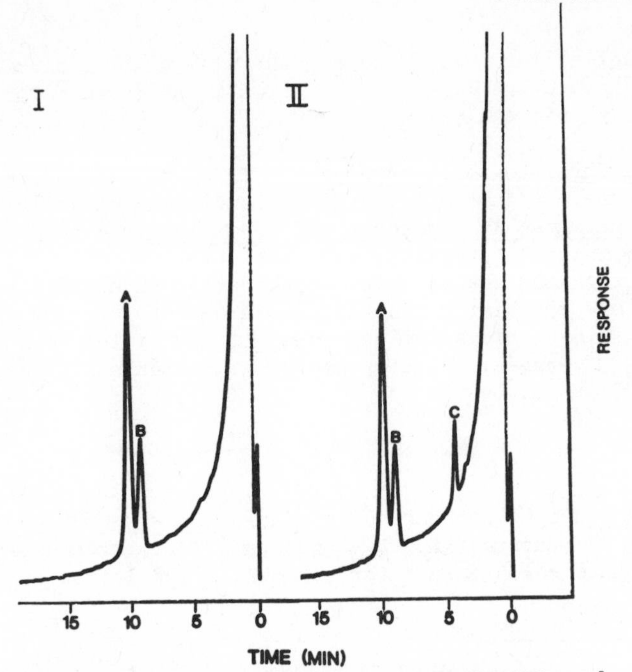

Figure 3. Gas–liquid chromatograms of FMC 33297 standard solution (35 μg/ml) before (I) *and after* (II) *cleanup:* A, trans *isomer peak;* B, cis *isomer peak;* C, *unknown reagent impurity peak*

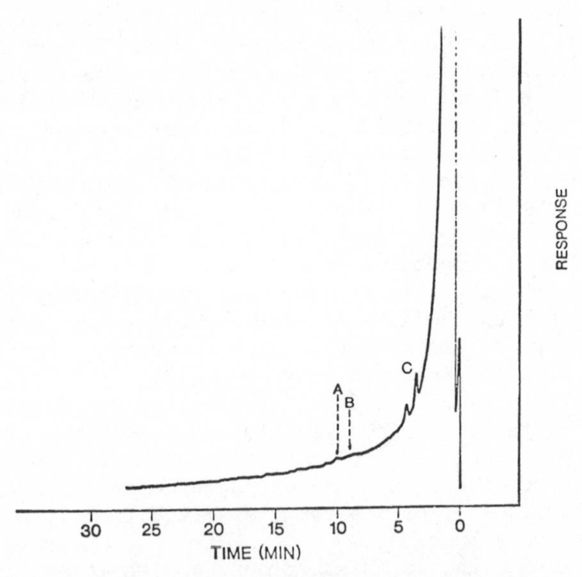

Figure 4. Gas–liquid chromatogram of FMC 33297 reagent blank: A, *retention time for* trans *isomer peak;* B, *retention time for* cis *isomer peak;* C, *impurity peaks introduced by reagents used for cleanup*

Table I. Average recoveries of FMC 33297 from fortified food commodities.

Added ppm	Corn		Cornmeal		Flour		Wheat	
	Found ppm	% recovery	Found ppm	% recovery	Found ppm	% recovery	Found ppm	% recovery
0.0	<0.2	---	<0.2	---	<0.2	---	<0.2	---
0.20	0.21	105	0.19	95.0	0.18	90.0	0.19	95.0
0.40	0.40	100	0.37	92.5	0.37	92.5	0.35	87.5
0.80	0.73	91.2	0.76	95.0	0.75	93.8	0.73	91.2
1.6	1.4	87.5	1.5	93.8	1.5	93.8	1.4	87.5
4.4	4.1	93.2	4.5	102.3	4.1	93.2	4.0	90.9
11.0	10.8	98.2	10.3	93.6	10.5	95.5	10.0	90.9
22.0	20.5	93.2	20.3	92.3	20.2	91.8	20.4	92.7

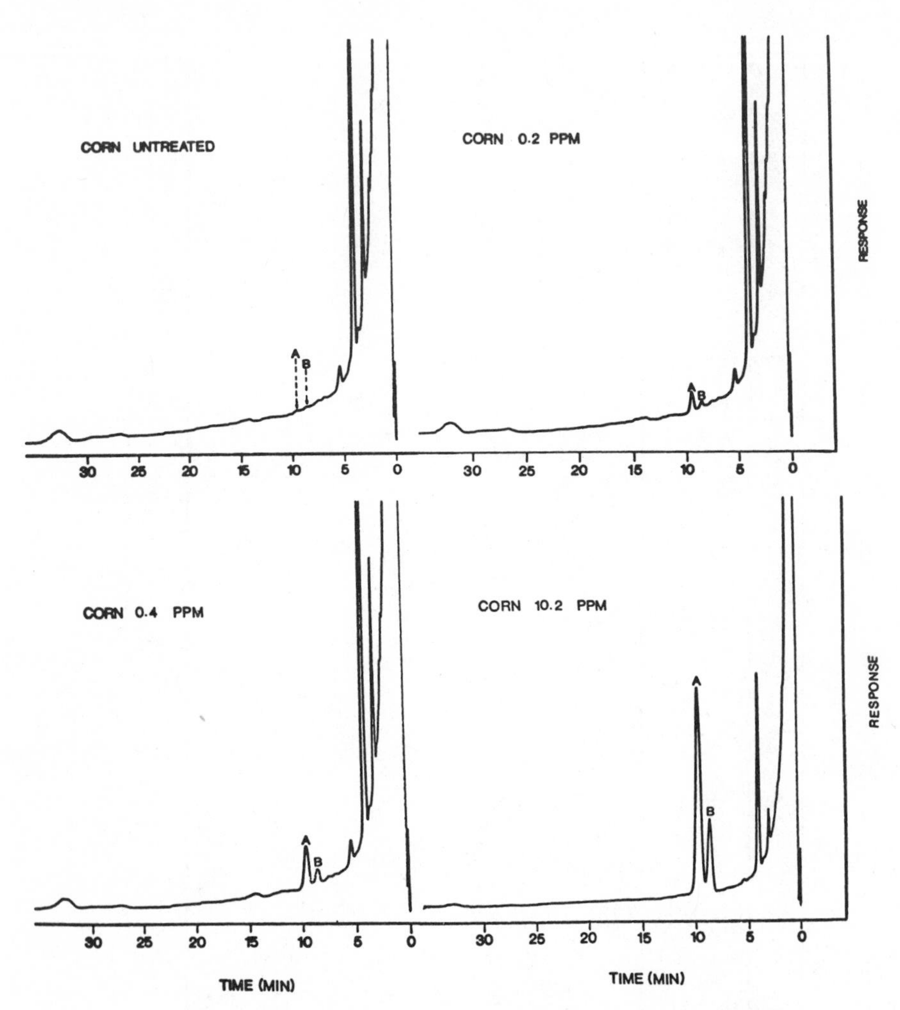

Figure 5. Gas–liquid chromatograms of corn commodity extracts unfortified and fortified with FMC 33297: A, trans *isomer peak;* B, cis *isomer peak*

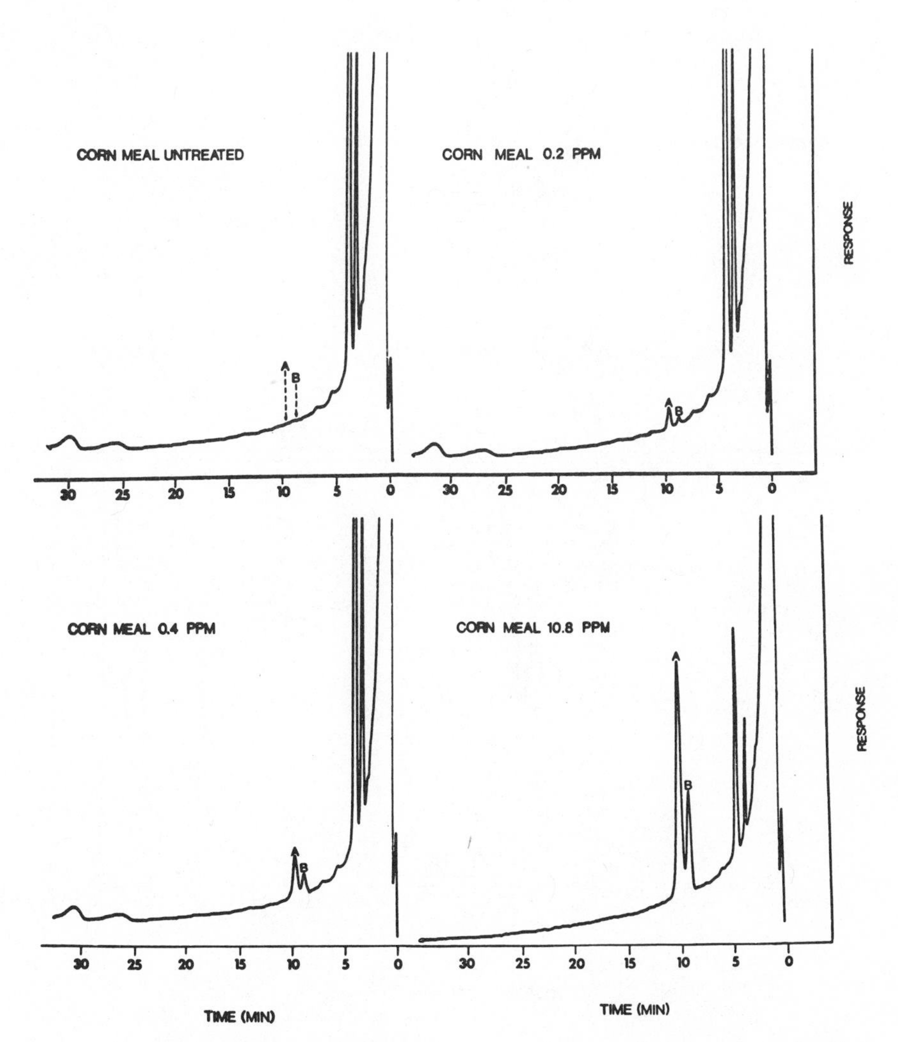

Figure 6. Gas–liquid chromatograms of corn meal commodity extracts unfortified and fortified with FMC 33297: A, trans *isomer peak;* B, cis *isomer peak*

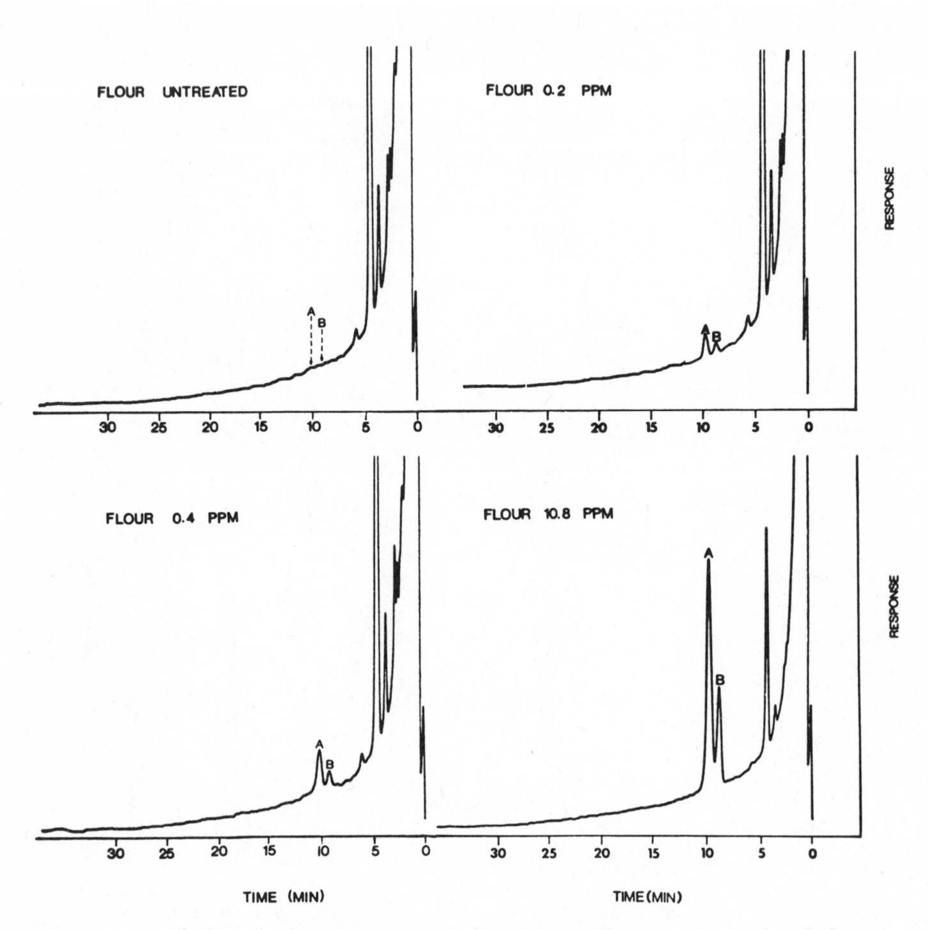

Figure 7. Gas–liquid chromatograms of flour commodity extracts unfortified and fortified with FMC 33297: A, trans *isomer peak;* B, cis *isomer peak*

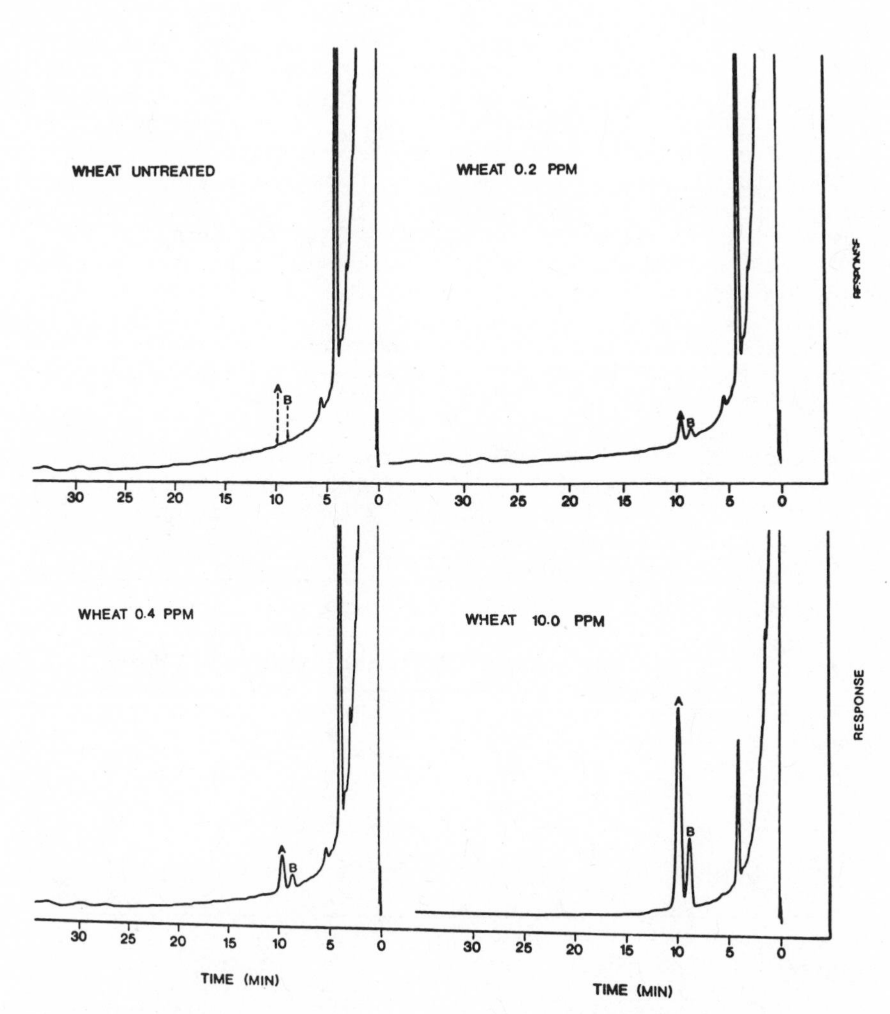

Figure 8. Gas–liquid chromatograms of wheat commodity extracts unfortified and fortified with FMC 33297: A, trans *isomer peak;* B, cis *isomer peak*

analyzed, as shown by results reported in Table I. Recoveries for separate determinations ranged from 88 to 105% for corn, 92 to 103% for cornmeal, 90 to 95% for flour, and 88 to 95% for wheat.

Chromatograms of the extracts of the commodities with and without FMC 33297 have been reported in Figures 5-8. The isomer peaks were fairly well resolved even at fortified concentration levels of 10 ppm. No interference peaks were observed at the FMC 33297 retention times for any of the commodities studied. Although the integrator was limited to the quantization level of 0.2 ppm, it was possible to use peak heights to obtain greater sensitivity. To calculate the sensitivity of the method it was found that peak height of the trans isomer could be used to give a minimum detectable concentration of 0.05 ppm based on twice the noise level.

Statistical analysis of the results for five samples of each commodity fortified at 0.5, 1.2, and 4.7 ppm was performed. The results are reported in Table II.

Table II. Standard deviations of a single determination for a series of five independent determinations on each commodity fortified at three treatment levels.

Fortification	Standard deviation for commodity			
level	Corn	Cornmeal	Flour	Wheat
(Ppm)				
0.60	0.030	0.019	0.020	0.029
1.2	0.092	0.020	0.031	0.018
4.7	0.070	0.061	0.049	0.14

Abstract

A simple and rapid gas-liquid chromatographic (GLC) method was developed for the determination of residues of the insecticide FMC 33297 (m-phenoxybenzyl cis,trans-(±)-3-(2,2-dichloro=vinyl)-2,2-dimethylcyclopropanecarboxylate), also known as NRDC 143, PP 557, and permethrin, in corn, cornmeal, flour, and wheat. The commodity was extracted with pentane and cleaned up by a solvent partition followed by liquid chromatography. The FMC 33297 residue was determined by GLC with a flame ionization detector. The results were compared with known standards that had undergone the same cleanup procedures. With electronic integration, the method was quantitative to concentrations of FMC 33297 to 0.2 ppm; with peak height measurement of the <u>trans</u> isomer peak, the method was sensitive to 0.05 ppm. Recoveries at levels of 0.20 to 22 ppm ranged from 88 to 105%. Reproducibility was good. The standard deviation for five determinations at levels of 0.60 to 4.7 ppm was 0.018 to 0.14% absolute for the four commodities.

Literature Cited

1. Bry, R. E., & Lang, J. H. J. Ga. Entomol. Soc. (1976) <u>11</u>, 4-9.
2. Davis, J. W., Harding, J. A., & Wolfenbarger, D. A. J. Econ. Entomol. (1975) <u>68</u>, 373-374.
3. Elliott, M., Farnham, A. W., Janes, N. F., Needham, P. H., Pulman, D. A., & Stevenson, J. H. Nature (1973) <u>246</u>, 169-170.
4. Berkovitch, I. Int. Pest Control (1974) <u>16</u>, 20.
5. Bry, R. E., Simonaitis, R. A., Lang, J. H., & Boatright, R. E. Soap/Cosmetics/Chem. Spec. (1976) <u>52</u>, 31-33, 98.

INDEX

D

E

F

G

H

O

P